CHINA ENGAGED

CHINA 2020 SERIES

China 2020:
Development Challenges in the New Century

Clear Water, Blue Skies:
China's Environment in the New Century

At China's Table:
Food Security Options

Financing Health Care:
Issues and Options for China

Sharing Rising Incomes:
Disparities in China

Old Age Security:
Pension Reform in China

China Engaged:
Integration with the Global Economy

THE WORLD BANK

WASHINGTON D.C.

CHINA 2020

CHINA
ENGAGED

INTEGRATION WITH THE GLOBAL ECONOMY

THE WORLD BANK
WASHINGTON D.C.

Copyright © 1997
The International Bank for Reconstruction
and Development/THE WORLD BANK
1818 H Street, N.W.
Washington, D.C. 20433, U.S.A.

Cover photograph by Serge Attal/ Gamma Liaison.

Cover insets (from left to right) by Vince Streano/Aristock, Inc.; Claus Meyer/ Black Star; Dennis Cox/China Stock; Dennis Cox/China Stock; Joe Carini/Pacific Stock; Erica Lansner/Black Star.

ISBN: 0-8213-4079-4

Contents

This report uses *Hong Kong* when referring to the Hong Kong
Special Administrative Region, People's Republic of China.

Acknowledgments

his report summarizes an internal World Bank paper entitled "China in the World Economy" (International Economics Department) and was prepared by a team led by Dipak Dasgupta. The other main authors were Milan Brahmbhatt, Caroline Farah, and Kim Murrell (chapter 1), Christian Bach and Will Martin (chapter 2), Kwang Jun and Xiaoqing Yu (chapter 3), and T.G. Srinivasan (chapter 4). The team is grateful to Nicholas Hope (former director of the China and Mongolia Department at the World Bank) for conceiving this study and for his generous support. Overall support and comments from Uri Dadush, Kathie Krumm, Vikram Nehru, and Richard Newfarmer are gratefully acknowledged.

The report also benefited from fruitful discussions with many Chinese government officials and analysts during a mission to China in January 1997. The mission team is thankful for the comments and information provided by a large number of Chinese officials, including Zhu Xian,

Zhao Jie, Wu Jinkang, Hu Xuehao, Xia Zhihua (Ministry of Finance); Cao Yushu, Hao Ju, Wang Xiduo, Wang Jianjun, Wu Qiang, Xiao Yanshun, Li Yunlin (State Planning Commission); Wang Zixian, Zhu Zhiping, Wang Zhihua, Chen Xin, Wang Yi, Deng Li, Ding Wei, Zhao Yanxia, Wang Jing, Chen Wenjing, Li Jian (Ministry of Foreign Trade and Economic Cooperation); Jing Xuecheng, Zhang Xinze, Yang Zaiping, Yao Keping, Yang Huisheng, Pei Chuanzhi, Zhao Xiangeng, Chen Xin, Liu Xuemei, Wan Cunzhi, Zhan Hongdi (People's Bank of China); Kang Qiang, Liu Ping, Jiang Hao, Chen Zhenchong, Dong Yufan, Lin Daqiang, Zhang Yimin, Jin Hongman (Customs General Administration); Wang Wei, Li Zhi, Yu Dayong (Customs Tariff Commission); Tu Guangshao (China Security Regulatory Commission); Song Yuzhong, Wei Dong, Wang Xu, Liu Dongsheng, Zhou Shuanghu (State Economic and Trade Commission); Chang Xiaochun, Jia Xiaozuo, Zhao Jian, Cheng Zhanyin, Guo Rongzi, Song Peiji, Shi Yonghong, Shen Qun (State Machinery and Electronics Import and Export Office); Huang Hongbo, Liu Xin, Xie Hemin, Xu Honglin, Hao Jinghua, Sun Qiumei, Zheng Haifeng, Fang Wen, Zhang Xiaopu (State Administration of Foreign Exchange); Lu Hailin, Ma Lin, Zhang Zhiyong, Hu Jinmu, Zhen Hua, Jin Donsheng, Yuan Yuan (State Tax Bureau); Chang Zhenmin (China International Trust and Investment Corporation); Zhu Min (Bank of China); Gao Fulai (China National Metals and Minerals Import and Export Corporation); and Chen Jiaying, Chen Boyu, Xue Jinglian, Song Jianhua, Xie Wei, Zhao Kunsheng, Lin Bifen, Chen Nenggeng, and Cao Dewang of Fujian Province. We would also like to thank all of those who participated in the seminar on the mission's initial findings, and who provided valuable advice and comments, including the helpful comments received on initial drafts of the study.

Assistance provided by Zhou Xiaobing and Li Sheng of the World Bank's Resident Mission in China and by Mei Hong, Pan Wenxin, and Fan Wenzhong of the Ministry of Finance is gratefully acknowledged.

The report was edited by Meta de Coquereaumont and Paul Holtz, laid out by Glenn McGrath and Laurel Morais, and designed by Kim Bieler, all with the American Writing Division of Communications Development Incorporated.

Overview

China was a closed economy until 1979. But as a result of open-door policies and reforms, its relationship with the rest of the world has been transformed and its modernization and growth have accelerated. As China's integration with the world economy progresses over the next twenty-five years, the implications will be enormous.

According to a projection scenario, China's share in world trade could more than triple to 10 percent, making it a major engine of growth for world trade. China would become the second largest trading nation in the world. The economic welfare benefits—for China and its trading partners—of China's liberalization and entry into the multilateral trading system would be large, with benefits to China alone estimated at more than $116 billion a year by 2005. Moreover, in a world of ever greater integration, benefits are expected in the form of rising wages in both industrial and developing countries. To realize the benefits, however, China and its major trading partners will need to further liberalize their trade and investment relationships within the context of a rules-based multilateral system, including China's accession to the World Trade Organization (WTO).

China in the world economy

The pace of global integration—the widening and intensifying of international linkages in trade, investment, and finance—has accelerated since the mid-1980s. China's opening to the world economy since 1978 has made major contributions to this process. The increase in China's trade to GDP ratio—from 10 percent in 1975–79 to 36 percent in 1990–94—was the seventh most rapid among 120 countries. And the increase in its foreign direct investment (FDI) to GDP ratio—from almost zero in 1975–79 to about 3.5 percent in 1990–94—was the sixth most rapid.

This greater openness in policies has paid rich dividends. China's faster and more efficient economic growth in the post-reform period is clearly associated with increased openness. Rapid growth in trade has accelerated economic growth. And foreign investment, attracted by the large pool of low-wage labor and a large and growing market, has encouraged faster specialization in labor-intensive manufacturing and increased employment and incomes, while reducing poverty. Even so, there remains much further potential for openness and for greater and more widely distributed productivity gains. China has essentially been catching up with other developing economies in opening to the world. Its 1990–94 its trade to GDP ratio was about the same as that of other large developing countries—but it was some 19 percentage points below the average for its East Asian neighbors. Again, China's FDI to GDP ratio is higher than that for other large developing countries, but compared with the East Asian average, not markedly so. Portfolio flows—equities, bonds, and other securities—to China were smaller (relative to GDP) than to other Asian economies or Latin America.

Foreign trade institutions and policies

China's foreign trade has become an engine of economic growth and an engine of growth in world trade.

The early phase of China's foreign trade reform encompassed the move from state monopoly to decentralized system. Special economic zones and open coastal cities led to greater openings for trade. Still, by the early 1990s China was falling behind other developing countries in liberalizing its trade. China's recent offers of trade reforms are positive, however, and promise to bring large benefits.

China's foreign trade institutions play an important role. But many are still oriented toward the direct control of trade, which will be increasingly incompatible with a liberalized trade system. Moreover, China's current trade policy regime has major inefficiencies. Reforms are needed to dismantle nontariff barriers, remove restraints on foreign trade, eliminate export restrictions, and move import policies toward a transparent system with low tariffs and no quotas.

Recent offers by China as part of its WTO accession process would move it far in that direction. The impact on industrial structure and overall welfare in China (and its main trading partners) would be very favorable. A general equilibrium trade model shows that over the next ten to fifteen years there would be large gains for the clothing sector (a result of abolishing the Multi-Fibre Arrangement), while reducing high import tariffs would bring a relative decline in the output of (capital- and technology-intensive) transport, machinery and equipment, and other heavy manufacturing industries (consistent with China's comparative advantage). Variable and relatively high rates of protection would remain for some sectors, however. Welfare gains from the proposed reforms would be large for China (equivalent to 4 percent higher annual output than under the baseline scenario) and significant for its major trading partners.

Integration with global capital markets

China's growing integration with global capital markets is reflected in its large shares in cross-border flows to developing countries: 40 percent of FDI flows, 10 percent of lending by banks, and growing portfolio flows.

Sustaining and shifting the composition of its FDI flows will be important for China in achieving some of its priority development objectives: more investment in infrastructure, faster technological progress, and greater flows of investment to interior and poorer provinces. To sustain FDI flows, China will need a stable macroeconomic environment and complementary reforms to improve the regulations and institutions governing FDI. And to shift the composition of these flows toward priorities, China will need to improve the transparency of its investment regime, which investors perceive to be one of the most complex in the world. Greater transparency in rules governing FDI and better property rights for investors will attract a wider range of FDI and ensure greater competition. This in turn would help ease the skewed geographical and sectoral distribution of FDI inflows, bring in and disseminate better technology, and attract more investment in infrastructure sectors. Trade policy reforms would also be important to bring FDI into more efficient areas and provide greater benefits to the economy.

Pressures in domestic financial markets to integrate faster with international financial markets are growing. Such integration also promises to bring about greater efficiency in domestic financial intermediation. In that context, domestic financial sector reform will be important. This would include the domestic banking and nonbank financial sectors. And reforms in capital markets will be essential, to improve efficiency and stability. China should move gradually, however, to integrate with global capital markets and to establish full convertibility of its capital account. Experience shows that international capital flows can be volatile and test governments' macroeconomic management. Weak financial and capital markets magnify these risks. Measures to improve commercial borrowing and debt management would also be prudent.

Global effects of China's integration to 2020

China's expected growth of about 7 percent a year over the next twenty-five years will magnify its presence in the world economy. China will account for some 40 percent of the increase in developing country imports between 1992 and 2020, helping to drive growth in world trade. The implications of China's growing role are favorable. Industrial countries will benefit from China's rising demand for imports of capital- and knowledge-intensive manufactures and services and primary products and from gains in terms of trade. Neighboring developing countries that are not close competitors (Asia's newly industrialized countries) are likely to experience equally significant gains.

For countries that are close competitors of China (low- and middle-income Asian countries such as India, Indonesia, the Philippines, and Thailand), world market shares and volumes of trade will likely continue to expand, but significant terms of trade losses are expected in their main exports of labor-intensive manufactures. Net income gains, however, will still be large. For developing countries and regions that do not directly trade much or compete with China (Latin America, Sub-Saharan Africa, Europe and Central Asia), there will be neither significant gains nor major losses. Finally, the greatest adverse effects of a slower integrating China would be in itself, although welfare losses for the rest of the world would also be significant.

chapter one

China in the World Economy

Starting from a position of near-autarky, China has been catching up rapidly with other developing countries in integrating with the global economy. Increased integration and openness have paid rich dividends in the way of faster growth. Still, there is much room for progress.

Integration with the world economy

A common indicator of economic integration is a country's participation in international markets (as measured by international trade and foreign capital flows relative to GDP, for example). Using trade openness to measure a country's integration with the world economy is not without problems, however, especially in the case of China. The country's size, problems in comparing its GDP with other countries', exchange rate movements, and the importance of its processing trade make comparisons difficult. A different approach is to define integration as the absence of policy barriers to the mobility of products and factors.

Trade openness

In 1975–79 China's international trade was 10 percent of GDP, the lowest among 120 countries studied.[1] By 1990–94 this ratio had risen to 36 percent, placing China in the top third. The increase in trade openness between the two periods—1.7 percentage points a year—was the seventh most rapid in the world.

But compared with its East Asian neighbors, the pace of China's integration was less impressive (figure 1.1). And in terms of level of integration China has essentially been catching up with other large developing countries (populations of more than 100 million), none of which (except Indonesia) had very open trade regimes. Consider China's trade to GDP ratio relative to that of other large developing countries —Bangladesh, Brazil, India, Indonesia, Nigeria, and Pakistan. In 1975–79 China's trade ratio was about 18 percentage points below the average for these countries; by 1990–94 it had risen to about the same level.

Country size and economy

Big countries tend to trade less across their borders than small ones because they have a wider range of resources and more opportunities for efficient specialization and trade within the country. Is China special in that regard?

A scatter diagram of population and trade to GDP ratios in 1990–94 suggests that China has done well, but it is not a major outlier (figure 1.2).

Discussions of China's trade openness also frequently turn to the size of its GDP, which is thought to be underestimated. Although it is true that developing country GDP can be underestimated because of the relatively low prices of nontradables, there are also data problems specific to China. The first issue can be addressed by using purchasing power parity–adjusted estimates of GDP, which correct for undervaluation of nontradables.[2] When applied to China, this brings the ratio of trade to GDP down to about 10 percent (in 1995), suggesting an economy more open than India (6 percent) but about the same as Brazil (10.5 percent). Nominal GDP is also thought to be underestimated. Thus China's high trade to GDP ratio, calculated conventionally, may be overestimating the openness of the economy.

Movements in the real exchange rate

Trade to GDP ratios are sensitive to movements in real exchange rates. Substantial depreciation of China's real official exchange rate (73 percent between 1980 and 1994) was associated with an increase in its trade ratio

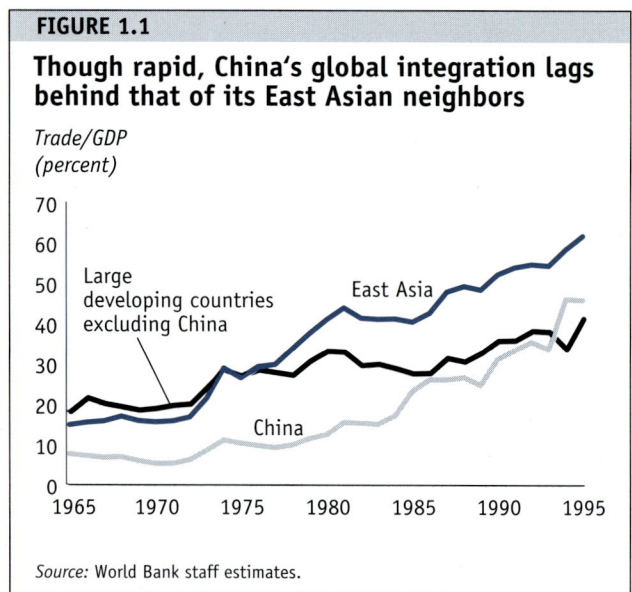

FIGURE 1.1

Though rapid, China's global integration lags behind that of its East Asian neighbors

Trade/GDP (percent)

Source: World Bank staff estimates.

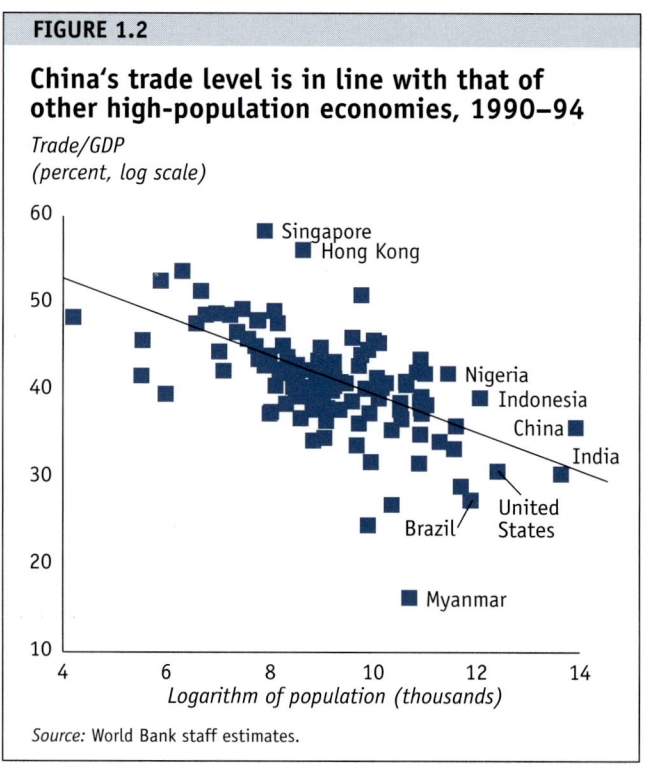

FIGURE 1.2

China's trade level is in line with that of other high-population economies, 1990–94

Trade/GDP (percent, log scale)

Source: World Bank staff estimates.

in the reform period. In constant prices the ratio rose 1.2 percentage points a year between 1975 and 1994, below the 1.7 points a year in nominal terms. This real rate of increase was the thirtieth most rapid among 120 countries studied, well below China's seventh place ranking in nominal terms.

A strong case can be made, however, that the fall in the real value of the yuan in the reform period was different from the fluctuations in the yen and the dollar in recent years and is a useful policy indicator of greater integration. Before reforms began, the exchange rate was a bookkeeping concept with little economic relevance. Extreme imbalances between supply and demand for foreign exchange were kept in check by tight exchange controls. Dismantling this system and moving toward a market-based system (marked by unification of the exchange rate system and real devaluation of the official exchange rate) was a precondition for global integration. Since these reforms were accomplished by 1994, the yuan has appreciated in real terms.

Processing and assembly trade

The increase in China's trade to GDP ratio was especially rapid after 1984, when the State Council approved two schemes allowing duty-free import of components and raw materials for use in export industries. Processed exports under these schemes rocketed to 45 percent of merchandise exports by 1991 and were more important in China's trade than were such products elsewhere in East Asia (Lardy 1994). The imported content of these processed goods averages 90 percent of export value (Lardy 1994, table 5.1; Fung 1996, table 14).

China, with its dependence on processed goods exports, will tend to stand higher in rankings of global integration based on the trade to GDP ratio than it would in rankings based on the share in GDP of value added that is exported or on the share of domestic consumption that is supplied by imports, both of which are better measures of integration. Still, the processing (or outsourcing) trade that has flourished in China will likely be of growing importance in world trade. Falling telecommunications and transportation costs allow a "slicing up of the value chain," enabling firms to locate various stages of production in countries with the appropriate comparative advantage (Krugman 1995). The second wave of newly industrialized countries in

TABLE 1.1

Share of goods sold at state-fixed prices, 1978–93
(percent)

Year	Retail commodities	Agricultural goods	Capital goods
1978	97	94	100
1992	10	15	20
1993	5	10	15

Source: Lardy 1994.

East Asia (Indonesia, Malaysia, Thailand) has been especially adept in attracting the labor-intensive stages of global and regional production. In China, however, high tariffs and deep exemptions create artificial incentives for this type of trade.

Trade policy

China's openness on trade policy was about the average for all large developing countries in the early 1990s. That average, however, was strongly influenced by South Asian countries, which traditionally have some of the world's highest levels of protection and have been among the slowest in trade liberalization. China's tariffs were substantially higher than those of other East Asian economies. Its nontariff barriers were around the middle of the range for all countries.

Prices

In a market economy the closeness of domestic to international prices of traded goods (expressed in a common currency) is often a preferred indicator of integration. In theory, these prices should differ only by the amount of trade barriers, transportation costs, and differences across countries in the cost of nontraded inputs.[3] Price data that are sufficiently detailed and extensive to allow such comparisons are rarely available, however.

World prices had almost no influence on prices in China before 1978, and there was a wide gap between domestic and international prices in the one year (1984) for which some comparative price data are available (Lardy 1992, 1994).[4] Price reforms since the mid-1980s mean that domestic prices for more than 90 percent of imported goods are now linked to (and affected by) world prices (table 1.1). This link may imply a narrowing of price differentials, although trade

Foreign direct investment in China has skyrocketed in recent years

Percentage of GDP

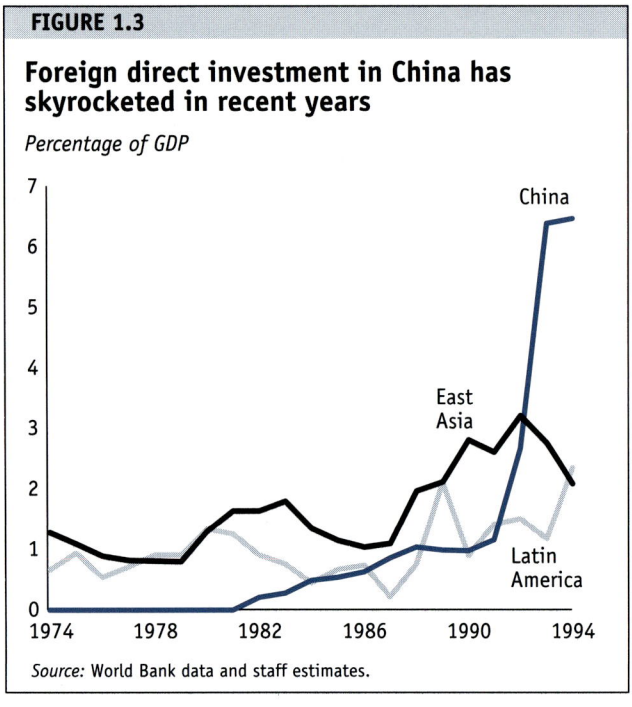

Source: World Bank data and staff estimates.

Portfolio capital flows to China remain limited

Percentage of GDP

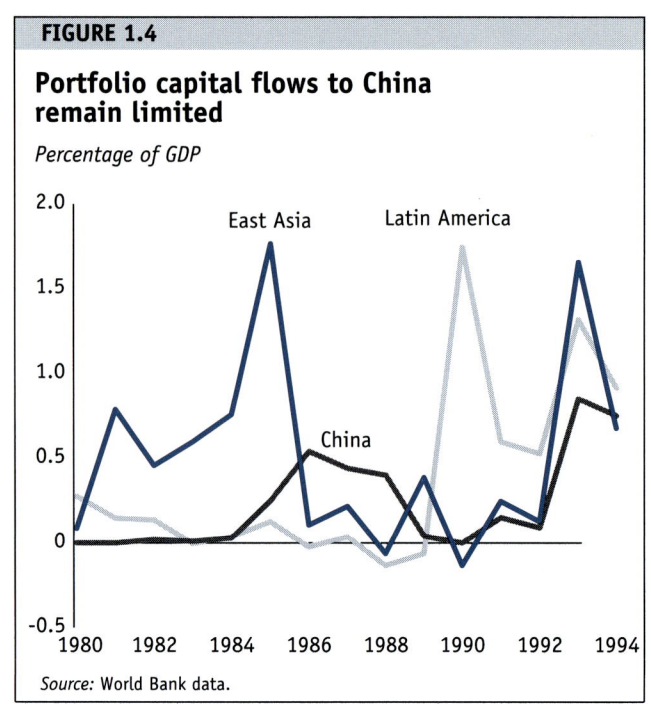

Source: World Bank data.

barriers and other costs continue to drive a wedge between the two.

Foreign direct investment

Inflows of foreign direct investment (FDI) to China, virtually nonexistent before 1979, rose to 3.5 percent of GDP in 1990–94 (figure 1.3).[5] In 1990–94 China's FDI to GDP ratio was the ninth highest among 120 countries studied. The increase between 1975 and 1994—0.24 percentage points a year—was the sixth highest. In 1996 FDI inflows rose to an estimated $42 billion.

Interpretation of cross-country data is more straightforward for FDI than for trade. There is no correlation between country size and FDI to GDP ratios. For China increased flows are in line with the steady liberalization of FDI regulations since 1979, especially in the coastal provinces. "Roundtripping," however, may account for some of the increase in the FDI to GDP ratio in recent years.[6] Moreover, FDI flows have been more volatile than trade. For many reasons, including the elimination of tax concessions for foreign investors in 1996 and the slowdown in the surge in transfers of labor-intensive assembly operations from neighboring countries, it is likely that high FDI inflows in China in 1993–95 (6.0–6.5 percent of GDP) were exceptional and will fall back to a level more sustainable in the long run.

Portfolio flows

During the 1980s portfolio capital flows were a phenomenon of high-income countries. Portfolio flows to developing countries picked up in the 1990s, but were concentrated. In only sixteen of seventy developing countries surveyed (not including China)—mostly in East Asia, Latin America, and developing Europe—did portfolio flows exceed 0.5 percent of GDP in 1990–94. In these sixteen countries, however, portfolio flows averaged about 2 percent of GDP; in the other countries flows averaged less than 0.1 percent of GDP. Continued restrictions and weak market infrastructure mean that portfolio flows to China have been limited. Such flows were negligible in 1980–84, rose to around 0.5 percent of GDP in 1986–88, and collapsed in 1989 because of political tensions. Flows revived modestly in the 1990s (to 0.7–0.9 percent of GDP in 1993–94) before slipping back in 1995 and 1996 (figure 1.4).

Openness and growth

There is a wealth of empirical evidence about the positive association between openness and economic growth. A study of a group of countries examined over 1970–89 found that, within the group of developing countries, open economies grew by 4.5 percent and closed ones by 0.7 percent (Sachs and Warner 1995). Among industrial countries, open economies grew at 2.3 percent and closed ones at 0.7 percent. Studies have also shown a link between faster integration and higher growth. Countries that have integrated fastest—that is, become more open in trade and investment—also achieved the most rapid growth in per capita GDP during 1983–94 (World Bank 1996).

Why do open and integrated economies grow faster? There are several reasons:

• *Allocative efficiency.* Higher levels of output are achieved when countries specialize according to their comparative advantage.

• *Factor accumulation and investment.* Trade liberalization accelerates investment by allowing access to bigger markets, permitting greater exploitation of increasing returns to scale, and encouraging imports of previously unavailable or cheaper capital goods (Murphy, Shleifer, and Vishny 1989). A recent cross-country study suggests that the effect of more open trade policy working through higher domestic investment may be the most important channel for growth (Wacziarg 1997). For China, rapid investment growth since 1978—due to increased openness—appears to have had a big effect on growth.

• *Knowledge spillovers.* Knowledge spillovers are likely in open economies because of the knowledge that is embodied in traded goods and machinery and FDI that is attracted to those economies.

• *Improved income distribution.* Returns to relatively abundant factors of production, such as labor in developing countries, will tend to be higher in open economies.

• *Government policy improvements.* There is greater pressure to maintain macroeconomic stability and smaller government in open economies.

Rapid growth has been driven by the same basic factors in China as it has in all East Asian economies: high rates of investment, financed by high domestic savings; increases in education spending; and abundant labor and its growth. In the post-reform period, market-driven investments and savings have responded to profitable investment opportunities, and human capital has been deployed more efficiently as labor became more responsive to market forces. Moreover, foreign trade and investment have played a bigger and more efficient allocative role in the economy.

Studies comparing pre- and post-reform growth suggest that China's economic growth accelerated after reforms started in 1979 (Borensztein and Ostry 1996). Remarkably, the improvement was achieved entirely by an improvement in total factor productivity growth— of between 3.6 and 4.1 percentage points.[7] A more recent study (World Bank 1997a) indicates that while the absolute size of the total factor productivity improvement may have been smaller (when corrected for data problems and differing methodological approaches), it still accounts for a large part (about a third) of the faster overall growth since reforms began.

A study using data for twenty-three industrial sectors in seven of China's coastal provinces concludes that four main factors contributed to faster growth, most related to greater openness:

• Open-door policies, particularly in special economic zones, were hugely successful in attracting more investment from overseas Chinese to the southeastern coast— as in the cases of Hong Kong (China) investors to Guangdong and, more recently, Taiwan (China) investors to Fujian. These provinces benefited not only from the physical proximity to overseas Chinese communities but also from their "relative backwardness" (Dollar 1992).

• Education has a positive effect on growth. There is an association between growth and the interaction between local education and foreign investment, which can be interpreted as a proxy for foreign knowledge and suggests a strong complementarity between the two. The finding that openness provides a crucial channel by which developing countries learn faster from industrial countries and the complementarities to education are also suggested in cross-country studies (Coe, Helpman, and Hoffmaister 1994).

• Physical infrastructure (and coastal locations) not only eases information flows but also provides the focal point for development of agglomerations (Shleifer 1991). This played a key role in China's growth.

• Growth, once initiated, has dynamic spillovers. On average, a 1 percentage point increase in growth of an industrial sector outside a region is associated with a 0.78 percentage point increase in growth of that industry within the region (Mody and Wang 1995).

Another study that examined export-led growth effects in China's cities found that exporting and foreign investment were strongly associated with faster industrial and economic growth (Wei 1996). In 1980–90 more exports were positively linked to higher industrial growth, whereas in the late 1980s cross-city growth differences were better explained by foreign investment.

Notes

1. Trade is defined as exports and imports of goods and nonfactor services as a share of GDP measured at market exchange rates.

2. For more details, see Summers and Heston (1991) and World Bank (1997b).

3. Evidence on the law of one price is summarized in Rogoff (1996), who notes, however, the persistence of large differences in prices even between countries (such as Canada and the United States) without trade barriers, with excellent transport links, and similar nontraded costs—a so-called border effect.

4. For thirty-nine individual goods domestic prices averaged 82 percent of world prices, with a standard deviation of 52 percent (Lardy 1992, table 4.2).

5. We normalize for GDP because, other things being equal, large economies draw more FDI than small ones.

6. Some sources estimate that up to 20 percent of recent inflows were domestic capital that was roundtripping to take advantage of incentives for foreign investors (UNCTAD 1995).

7. Borensztein and Ostry (1996) note that the estimate for total factor productivity growth may be overestimated by 1 percentage point because of the exclusion of human capital from the growth-accounting exercise.

References

Borensztein, Eduardo, and Jonathan Ostry. 1996. "Accounting for China's Growth Performance." *American Economic Review Papers and Proceedings* 86(2): 224–28.

Coe, David, Elhanan Helpman, and Alexander Hoffmaister. 1994. "North-South R&D Spillovers." IMF Working Paper WP/94/144. International Monetary Fund, Washington, D.C.

Dasgupta, Dipak, 1993. "Why Some Regions Do Better Than Others." In Amex Bank, *Finance and the International Economy.* New York: Oxford University Press.

Dollar, David, 1992. "Exploiting the Advantages of Backwardness: The Importance of Education and Outward Orientation." World Bank, Policy Research Department, Washington, D.C.

Fung, K.C. 1996. "Accounting for Chinese Trade: Some National and Regional Considerations." NBER Working Paper 5595. National Bureau for Economic Research, Cambridge, Mass.

Krugman, Paul. 1995. "Growing World Trade: Causes and Consequences." *Brookings Papers on Economic Activity 1.* Brookings Institution, Washington, D.C.

Lardy, Nicholas. 1992. *Foreign Trade and Economic Reform in China, 1978–1990.* Cambridge: Cambridge University Press.

———. 1994. *China in the World Economy.* Washington, D.C.: Institute for International Economics.

Lardy, Nicholas, and Fritz Machlup. 1977. *A History of Thought on Economic Integration.* New York: Columbia University Press.

Mody, Ashoka, and Fang-Yi Wang. 1995. "Explaining Industrial Growth in Coastal China." PSD Occasional Paper. World Bank, Washington, D.C.

Murphy, Kevin, Andrei Shleifer, and Robert Vishny. 1989. "Industrialization and the Big Push." *Journal of Political Economy* 97 (5).

Ng, Francis, and Alexander Yeats. 1996. "Open Economies Work Better! Did Africa's Protectionist Policies Cause Its Marginalization in World Trade?" Policy Research Working Paper 1636. World Bank, Washington, D.C.

Rogoff, Kenneth. 1996. "The Purchasing Power Parity Puzzle." *Journal of Economic Literature* 34 (June): 647–68.

Sachs, Jeffrey, and Andrew Warner. 1995. "Economic Reform and the Process of Global Integration." *Brookings Papers on Economic Activity 1.* Brookings Institution, Washington, D.C.

Shang, Jin Wei. 1993. "Open Door Policy and China's Rapid Growth: Evidence from City-Level Data." NBER Working Paper 4602. National Bureau of Economic Research, Cambridge, Mass.

Shleifer, Andrei. 1991. "Externalities and Economic Growth: Lessons from Recent Work." Background paper for *World Development Report 1991.* World Bank, World Development Report Office, Washington, D.C.

Summers, Robert, and Alan Heston. 1991. "Penn World Tables (Mark 5): An Expanded Set of International Comparisons, 1950–1988." *Quarterly Journal of Economics* 106(May): 327–68.

UNCTAD (United Nations Conference on Trade and Development). 1995. *World Investment Report 1995.* Geneva.

Wacziarg, R. 1997. "Measuring the Dynamic Gains from Trade." Harvard University, Department of Economics, Cambridge, Mass.

Wei, S. 1996. "Foreign Direct Investment in China." In T. Ito and A. Krueger, eds., *Financial Deregulation and Integration in East Asia.* University of Chicago Press.

World Bank. 1996. *Global Economic Prospects and the Developing Countries 1996.* Washington, D.C.

———. 1997a. *China 2020: Development Challenges in the New Century.* Washington, D.C.

———. 1997b. *World Development Indicators 1997.* Washington, D.C.

chapter two

Foreign Trade Institutions and Policies

China's foreign trade institutions play an important role in setting and implementing policies. Although enormous progress has been made, some of these institutions remain oriented toward direct control of trade, which will be incompatible with a liberalized trade regime. Moreover, China's trade regime has major inefficiencies. Reforms are needed to dismantle the maze of nontariff barriers, remove restraints on foreign trading, eliminate export restrictions, move import policies toward a transparent system with low tariff rates, and abolish import quotas.

Recent offers by China to reduce tariff and nontariff barriers would help greatly in making better use of its comparative advantage and increasing its openness. They would also reduce the strong incentives to focus on processing trade that are inherent in the current system of high tariffs and deep exemptions. Still, high and varying rates of protection would remain for a number of manufacturing industries. The impact of such reforms on the industrial structure and overall welfare in China (and its main trading partners) would be very favorable. A general equilibrium model shows that there would be dramatic gains for the clothing sector (a result of abolishing the Multi-Fibre Arrangement), while reducing high protection would stimulate efficient growth in the labor-intensive apparel sector and the capital- and skill-intensive transport equipment sector.

Foreign trade institutions and trade barriers

Eight institutions are involved in formulating or administering China's trade policies—the Ministry of Foreign Trade and Economic Cooperation, State Planning Commission, State Economic and Trade Commission, Customs Tariff Commission, Customs General Administration, National Machinery and Electronics Import and Export Office, State Administration of Foreign Exchange (under the People's Bank of China), and foreign trade corporations. These agencies administer a set of complex and frequently overlapping controls over foreign trade (figure 2.1). China controls the entry of enterprises into foreign trade (through foreign trade rights) as well as trade itself.

Foreign trade rights

Before reform, foreign trade in China was monopolized by ten to sixteen specialized foreign trade corporations (Lardy 1992). Since then, there has been a dramatic increase in the number of firms allowed to conduct foreign trade. In 1994 the number of foreign trade corporations engaged in exporting was estimated at 9,400 and those in importing at 8,700.[1] Thousands of big state enterprises now have direct trading rights. Foreign-affiliated enterprises (including joint ventures and foreign-owned firms) also have rights to trade and are exempt from licensing on their products and inputs (MOFTEC 1994, p. 155). In 1994 joint ventures accounted for a third of China's imports and almost 20 percent of exports (table 2.1). Foreign-owned enterprises accounted for 12 percent of imports and 9 percent of exports. Given the large number of participants in foreign trade, there seems little reason for China to continue restricting access to trading rights. Such policies have already been abolished in Russia and all of the transition economies of Eastern Europe.

Export controls

China's export taxes and controls appear to have caused less economic damage than might be expected.

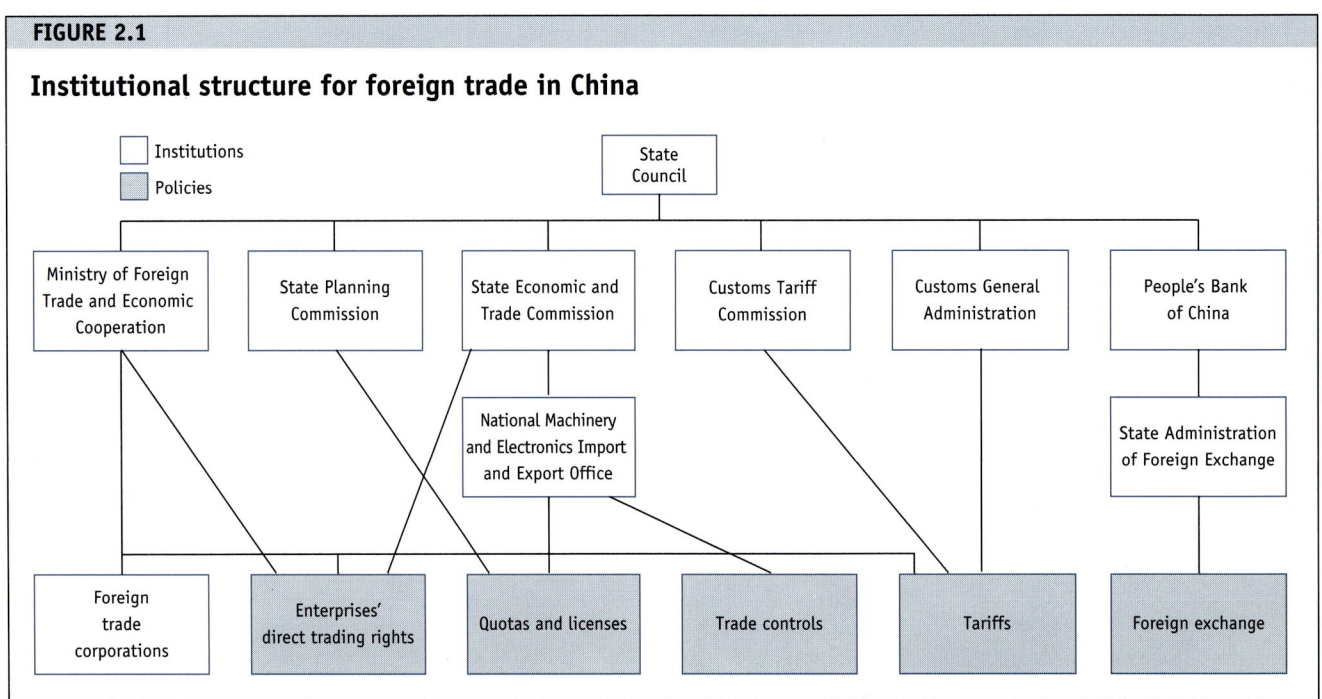

FIGURE 2.1

Institutional structure for foreign trade in China

TABLE 2.1
Contribution of different firms to China's trade, 1994

	Exporting		Importing	
Type of firm	Number	Share of exports (percent)	Number	Share of imports (percent)
Foreign trade corporations	9,400	53	8,700	44
State-owned enterprises	7,800	17	3,600	8
Joint ventures	30,000	19	64,800	34
Foreign-owned firms	9,730	9	23,239	12
Collective and private firms	1,060	1	1,828	1
Other	520	0.2	5,378	1
Total	58,500	100	107,513	100

Note: Some numbers may not add because of rounding.
Source: ITC 1995, p. 22.

They lead to rent-seeking, however, and make it difficult for new exporters to enter some markets. Export taxes are used to improve terms of trade for a few commodities in which China has a large market share or for which it wants to influence domestic prices. Export licenses are issued for goods subject to price coordination—that is, prices determined by the Chamber of Commerce. The traditional means of allocating export quotas has been administrative, although some are being allocated through bidding procedures on an experimental basis (Editorial Board of the Almanac of Chinese Foreign Economic Relations and Trade 1995).

Import controls

China's import policies include various tariff and nontariff measures that result in high protection and deep exemptions. The efficiency and performance of the economy could be improved substantially by moving to lower and more uniform protection.

Tariff exemptions have been widely granted for inputs used in exports, capital goods used in export processing zones, and for technological upgrading. According to the Customs General Administration, exemptions for export processing covered 44 percent of merchandise imports in 1995 and 45 percent in 1996. As a result tariff collections have fallen well below the statutory rate. In 1996 the average tariff was 19.8 percent, and a value-added tax (VAT) of 17 percent applied on most commodities. But collections of the VAT and customs duties were only 7.4 percent for all imports, or

13.4 percent on imports other than those for export processing. In 1994 exemptions for capital goods imports started being phased out. Exemptions for capital goods used by foreign-funded enterprises and for technical upgrading were abolished in 1996, subject to a one- to two-year grandfathering provision not extending beyond 1997.

There are quotas for twenty-six general commodities for which it is feared that excess imports could bring serious injury to the development of domestic industry or damage to China's balance of payments. The State Planning Commission determines the quota levels and allocates them to provinces. There are quotas on fifteen machinery and electronic products (for example, automobiles and refrigerators).

Automatic import registration covers a wide range of imports, including oil, steel, copper, nonferrous metals, wool, polyester, tobacco, and cotton. Some of these products are also subject to state trading, some to designated trading, and some to quotas and licenses, while some are otherwise unrestricted. While automatic import registration is not intended as a nontariff barrier, it could have this effect because it is contingent on a market-need criterion (box 2.1).

China's VAT on imports and domestically produced goods is 17 percent (13 percent on selected goods). Exports are zero-rated and eligible for a rebate on the VAT paid on inputs. The claims for rebates substantially exceeded expectations, however, and in 1995 the rebate on inputs purchased for export production was reduced to 14 percent and later to 9 percent. Thus the VAT imposes an effective net tax on exports that comply with the law. Improvements in the administration of the law that would raise the collection rate and allow a return to zero rating for exports would substantially improve trade performance.

China's trade reforms

During the 1980s and early 1990s developing countries began to lower protection rates. China, however, kept its average tariff at roughly the level of the early 1980s (table 2.2). By the early 1990s China's average tariffs were among the highest in the world, exceeded only by those in Egypt, India, Pakistan, and Thailand of the comparator countries considered. Since then, there has been a wave of further liberalization in developing

Steel imports and trade barriers

Steel is considered essential to China's economy and so has been subjected to import quotas. Following a Memorandum of Understanding on Market Access with the United States, China agreed to remove quotas by the end of 1993. Automatic import registration was introduced in 1994 to monitor steel imports. However, registration requires several conditions, including that imports meet a market need. At the national level, three state foreign trade corporations have exclusive rights to import steel. At the provincial level, fifty-five companies have the right to import steel to meet local needs. In addition, foreign enterprises have rights to import steel for their own use.

In recent years steel imports have fluctuated dramatically. In 1995 they were less than half of 1993 (see table). Why? The dramatic increase in steel imports in 1993, still under quota administration, was partly due to strong demand by foreign and processing enterprises; the price difference between world and domestic markets drove those enterprises to increase steel imports, which they sometimes resold on the domestic market at a profit. The later decline in steel imports appears to have been influenced by the automatic import registration system, together with restrictions on steel imported through special economic zones and foreign-invested enterprises.

Imports of finished steel, 1990–96
(millions of tons)

	1990	1991	1992	1993	1994	1995	1996
Imports	3.7	3.3	6.2	30.5	22.8	14.0	16.0

Source: Dickson 1996; China Daily Business Weekly, December 1–7, 1996.

Weighted average most-favored-nation tariffs for large developing countries, 1980–93
(percent)

Country	1980–83	1984–87	1988–90	1991–93	Post–Uruguay round
Brazil	—	50.2	28.4	14.7	11.7
China	31.9	29.2	29.2	30.6	16.6[a]
Egypt	30.4	30.6	22.9	31.0	—
India	59.9	90.0	62.4	42.6	30.9
Indonesia	23.5	18.2	18.0	12.6	10.7
Korea, Rep. of	—	20.2	11.3	10.0	7.7
Malaysia	9.7	14.7	11.5	11.2	6.4
Mexico	23.6	9.1	8.9	12.3	10.4
Pakistan	57.7	59.7	40.1	56.2	—
Thailand	24.8	26.9	38.0	36.9	26.1
Turkey	—	21.9	19.0	9.0	2.8[b]

a. Takes into account the effects of China's 1996 tariff reductions and the effects of the most recent World Trade Organization accession offer.
b. Reflects Turkey's entry into a free trade agreement with the European Union.
Source: UNCTAD 1994; Finger, Ingco, and Reincke 1996; Harrison, Rutherford, and Tarr 1996.

rate a strong top-down element, which is expected to reduce considerably the variability of the tariff regime; the standard deviation of the tariff schedule is expected to fall by 58 percent under the proposals.

Effective rates of protection

The effective rate of protection provides a measure of the incentives for expanding or contracting certain activities (Corden 1971). China has high and variable effective rates of protection in various manufacturing industries (figure 2.2). Prospective reforms will lead to substantial reductions in effective protection, although rates will remain high in most heavy industries.

Nontariff barriers

Nontariff barriers applied include state monopoly trading, designated trading, the London Convention on Chemical Products, import licensing, quotas, and price tendering (table 2.3). These categories cover almost a third of China's imports (down from more than half in 1992). The nontariff barrier with the greatest coverage is import licensing—more than 18 percent. State trading is the next biggest category, at 11 percent, while designated trading covers more than 7 percent of imports. Import tendering also covers more than 7 percent, exclusively in machinery and transport. The tariff

countries. China's tariff reductions to date and offered reductions under the accession process for the World Trade Organization (WTO) promise to bring it more in line with other large developing economies.

China has committed itself to further trade liberalization and offered to reduce tariffs and eliminate most nontariff barriers (see annex). Reductions in applied tariff rates announced at the November 1995 meeting of the Asia-Pacific Economic Cooperation group (effective in 1996) are selective. Tariffs fall sharply on coal and gas, textiles, apparel and leather, and heavy manufactures such as transport equipment, fabricated metal products, and machinery. These cuts reduce the trade-weighted average tariff from 28.1 percent to 19.8 percent and the unweighted average from 36 percent to 23 percent. Under China's current WTO accession proposal, the estimated average tariff in 2005 would fall to 16.2 percent. The proposed tariff reforms also incorpo-

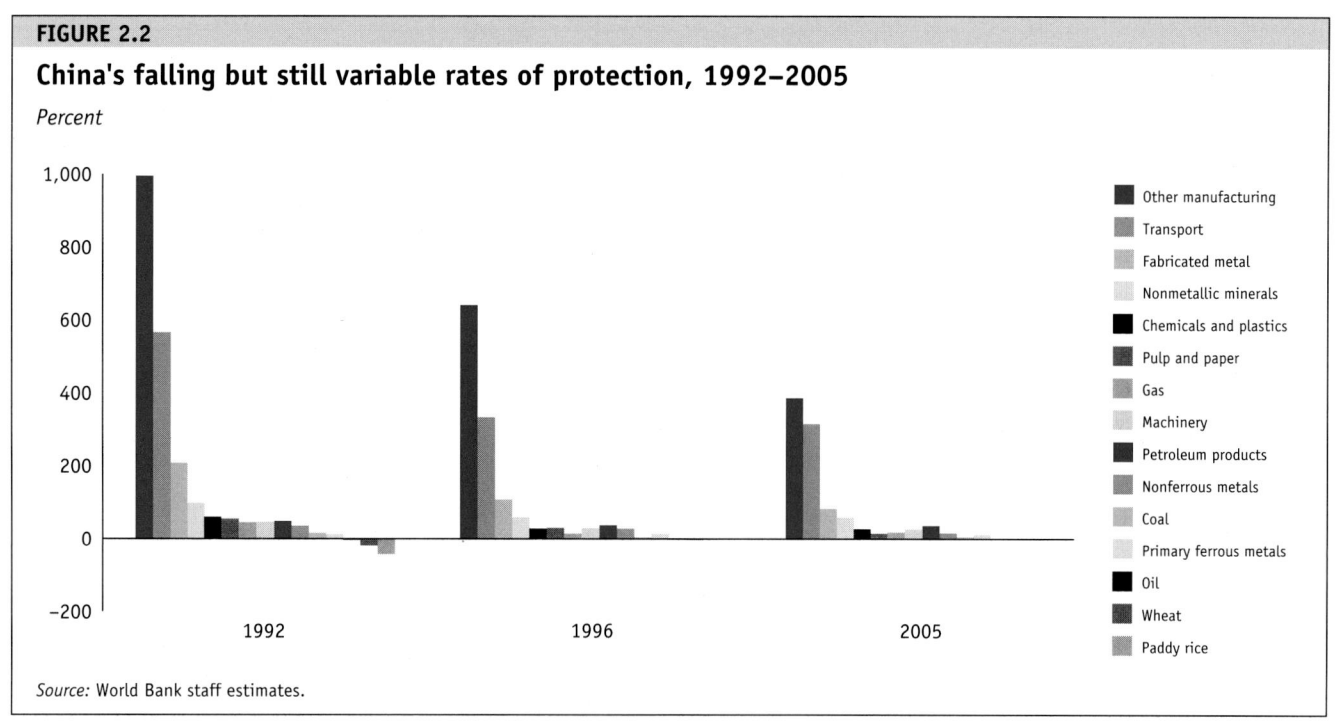

FIGURE 2.2

China's falling but still variable rates of protection, 1992–2005

Percent

Legend:
- Other manufacturing
- Transport
- Fabricated metal
- Nonmetallic minerals
- Chemicals and plastics
- Pulp and paper
- Gas
- Machinery
- Petroleum products
- Nonferrous metals
- Coal
- Primary ferrous metals
- Oil
- Wheat
- Paddy rice

X-axis: 1992, 1996, 2005

Source: World Bank staff estimates.

TABLE 2.3

Nontariff measures affecting China's imports in 1996
(percent coverage)

Import	State trading	Designated trading	London Convention on Chemical Products	Import licensing	Quotas	Price tendering	All
Rice	100.0	0.0	0.0	100.0	0.0	0.0	100.0
Wheat	100.0	0.0	0.0	100.0	0.0	0.0	100.0
Coarse grains	0.0	0.0	0.0	0.0	0.0	0.0	0.0
Nongrain crops	50.0	22.9	0.0	72.9	72.9	0.0	72.9
Livestock	0.0	72.7	0.0	72.7	72.7	0.0	72.7
Meat and milk	0.0	0.0	0.0	0.0	0.0	0.0	0.0
Other food products	37.2	0.0	0.0	32.9	31.7	0.0	38.4
Natural resources	46.6	12.8	0.0	0.0	0.0	0.0	59.5
Textiles	0.3	5.7	0.0	12.7	12.7	0.0	12.7
Wearing apparel	0.0	0.0	0.0	0.0	0.0	0.0	0.0
Light manufactures	0.0	9.3	0.0	0.0	0.0	0.0	9.3
Transport industries	0.0	0.0	0.0	35.8	35.8	6.6	42.4
Machinery and equipment	0.0	0.0	0.0	9.2	9.2	20.4	26.8
Basic heavy manufactures	18.7	16.2	0.3	23.5	22.7	0.0	37.7
Services	0.0	0.0	0.0	0.0	0.0	0.0	0.0
Total	11.0	7.3	0.1	18.5	16.3	7.4	32.5

Note: Using 1992 trade weights.
Source: World Bank staff calculations using UNCTAD data and WTO 1996 data.

equivalent of China's nontariff barriers (drawing on International Comparison Project data) is estimated to average 9.3 percent for 1996. The phaseout of nontariff barriers contained in China's WTO offer (see below) would reduce this to nearly zero.

As part of its negotiations for accession, China has committed to eliminating most nontariff barriers, except for state trading, which is covered by special provisions under the WTO. Immediately on joining the WTO, coverage of nontariff barriers would fall dramatically. After full phase-in, the only significant remaining measure would be state trading (mainly in agriculture), and overall nontariff barrier coverage would fall to 11 percent.

TABLE 2.4

Changes in industrial output under different trade simulations

(percent)

Sector	Baseline 1992–2005	Change from baseline				
		UR	C96	WTO	CTW	NTM
Rice	39	0	1	1	2	1
Wheat	20	5	4	3	1	1
Coarse grains	47	3	5	4	6	6
Nongrain crops	74	0	4	2	4	4
Livestock	113	2	9	11	18	20
Meat and milk	80	13	20	19	17	20
Other food products	88	−6	−11	−22	−25	−32
Natural resources	248	4	4	8	−15	−8
Textiles	134	9	−18	−17	35	37
Wearing apparel	142	4	88	103	469	494
Light manufactures	177	2	19	22	2	16
Transport equipment	773	−17	274	482	273	292
Machinery	354	−5	−2	5	−23	−44
Other heavy manufactures	279	1	5	9	−7	1
Services	224	−1	4	6	6	8
Capital goods	217	1	2	2	2	3

Note: For explanation of variables, see annex table A.4
Source: World Bank staff estimates.

TABLE 2.5

Changes in welfare under different trade simulations

(percent)

Economy	Baseline 1992–2005	Change from baseline				
		UR	C96	WTO	CTW	NTM
China	138	0.4	5.9	7.1	10.6	12.4
Hong Kong and Taiwan, China	81	0.2	1.7	2.2	2.3	2.7
Korea, Rep. of, and Singapore	104	8.9	9.4	9.7	9.5	9.6
Rest of ASEAN[a]	94	8.8	8.9	9.0	6.6	6.6
South Asia	51	2.1	2.1	2.1	1.2	0.9
Japan	35	1.0	1.2	1.3	1.3	1.4
Australia and New Zealand	34	0.6	0.7	0.7	0.8	0.9
NAFTA[b]	22	0.4	0.4	0.4	0.5	0.5
Western Europe	35	0.7	0.7	0.7	0.8	0.8
Rest of world	15	0.0	0.0	0.0	0.0	0.1

Note: For explanation of variables, see annex table A.4
a. Association of Southeast Asian Nations.
b. Canada, Mexico, and the United States.
Source: World Bank staff estimates.

Liberalization proposals

The implications of China's offers of tariff reductions were examined using a general equilibrium trade model of the world economy (Hertel 1997). The design of the simulations used to analyze these reforms is set out in annex table A.4. The proposed reductions are estimated to have substantial effects on China's industrial output and welfare (tables 2.4 and 2.5). The first four scenarios (UR, C96, WTO, and CTW) deal only with the effects of China's tariff liberalization, while the final scenario (NTM) adds the estimated impact of the proposed phaseout of nontariff barriers.

The dynamic growth of the Chinese economy results in rapid expansion of capital- and skill-intensive sectors, such as transport equipment and machinery, relative to agriculture and food products. The 1996 liberalization package is expected to stimulate the wearing apparel and transport equipment sectors by lowering their costs and expanding exports, although expansion of apparel output is constrained by Multifibre Arrangement (MFA) quotas in the major markets. Implementing China's WTO tariff offer would further stimulate these two sectors. Following this, abolition of China's MFA quotas would lead to a dramatic

expansion of the apparel sector. Given China's strong comparative advantage in these labor-intensive goods, and the strong repression of the sector by the MFA, the output of apparel would expand by close to 500 percent, with the output of transport equipment still growing by more than 230 percent. Abolishing nontariff measures (NTMs) reinforces the changes in the structure of output generated by the tariff reductions. It should be noted that the reforms considered above would not cause output of any industry to decline. They have their impact on industry structure primarily by channeling new resources into the sectors where the returns to the Chinese economy are greatest.

Moreover, there will be substantial welfare benefits to China from reforms (see table 2.5). The 1996 reductions generate significant gains, as do the tariff reductions associated with the introduction of tariff bindings under WTO offers. The largest incremental gain, however, comes from the elimination of the MFA quotas in the CTW simulation, particularly when this is augmented by the proposed elimination of nontariff barriers under the NTM scenario.

Note

1. This estimate, from ITC (1995), was based on the firm codes reported to the Customs General Administration, which distinguish

between subsidiaries of the foreign trade corporations authorized by the Ministry of Foreign Trade and Economic Cooperation.

References

Anderson, J. 1995. "Tariff Index Theory." *Review of International Economics* 3(2): 156–74.

Bach, C., W. Martin, and J. Stevens. 1996. "China and the WTO: Tariff Offers, Exemptions and Welfare Implications." *Weltwirtschaftliches Archiv* 132(3): 409–31.

Corden, W.M. 1971. *The Theory of Protection.* Oxford: Oxford University Press.

Dickson, Ian. 1996. "China's Steel Imports: An Outline of Recent Trade Barriers." Working Paper 96/6. University of Adelaide, Chinese Economy Research Unit, Australia.

Editorial Board of the Almanac of China's Foreign Economic Relations and Trade. 1995. *Almanac of China's Foreign Economic Relations and Trade (1994/1995).* China Resources Advertising Co., Ltd.

Fan, Baoqing. 1995. "The Reform of the Foreign Trade Administration System." *Great Economic Reforms Series.* Beijing: Zhongguo Jihua Chubanshe.

Finger, M. 1996. "Legalized Backsliding: Safeguard Provisions in GATT." In W. Martin and L.A. Winters, eds., *The Uruguay Round and the Developing Countries.* Cambridge: Cambridge University Press.

Finger, J.M., M. Ingco, and U. Reincke. 1996. *The Uruguay Round: Statistics on Tariff Concessions Given and Received.* Washington, D.C.: World Bank.

Francois, J., and W. Martin. 1995. "Multilateral Trade Rules and the Cost of Protection." Discussion Paper 1214. Centre for Economic Policy Research, London.

Gros, D. 1994. "Comment on Russian Trade Policy." In C. Michalopoulos and D. Tarr, eds., *Trade in the New Independent States.* Studies of Economies in Transformation 13. Washington, D.C.: World Bank.

Harrison, G., T. Rutherford, and D. Tarr. 1996. "Economic Implications for Turkey of a Customs Union with the European Union." Policy Research Working Paper 1599. World Bank, Washington, D.C.

Hertel, Thomas, ed. 1997. *Global Trade Analysis: Modeling and Applications.* Cambridge: Cambridge University Press.

Hsiao, G.T. 1977. *Foreign Trade of China: Policy, Law, and Practice.* Berkeley: University of California Press.

Hsu, John C. 1989. *China's Foreign Trade Reforms.* Cambridge: Cambridge University Press.

ITC (International Trade Centre). 1995. *Survey of China's Foreign Trade: An Analysis of China's Export and Import Data at the Enterprise Level.* Geneva.

Krugman, Paul. 1993. "Protection in Developing Countries." In Rudiger Dornbusch, ed., *Policymaking in the Open Economy: Concepts and Case Studies in Economic Performance.* New York: Oxford University Press.

Kueh, Y.Y. 1987. "Economic decentralization and foreign trade expansion in China." In J. Chai and C.K. Leung, eds., *China's Economic Reforms.* University of Hong Kong, Centre of Asian Studies.

Lardy, Nicholas. 1992. *Foreign Trade and Economic Reform in China, 1978–1990.* Cambridge: Cambridge University Press.

———. 1995. "The Role of Foreign Trade and Investment in China's Economic Transformation." *The China Quarterly* (December).

MOFTEC (Ministry of Foreign Economic Cooperation). 1994. *Collection of Laws and Regulations of the People's Republic of China Concerning Foreign Economic Relations.* Beijing: Department of Treaty and Law.

———. 1996. "Provisional Measures on the Establishment of Sino-Foreign Joint Venture Trading Companies on a Pilot Basis." Decree 3. Beijing.

Sun, Xiao. 1997. "Foreign Trade JVs to Boost Exports." *China Daily Business Weekly* 12 (January).

UNCTAD (United Nations Conference on Trade and Development). 1994. *Directory of Import Regimes, Part I: Monitoring Import Regimes.* New York.

Wall, D. 1996. "China's Special Economic Zones." In D. Wall, Jiang Boke, and Xiangshuo Yin, eds., *China's Opening Door.* London: Royal Institute of International Affairs.

Wang, Hong. 1993. *China's Exports since 1978.* New York: St. Martin's Press.

World Bank. 1994. *China: Foreign Trade Reform.* Washington, D.C.

WTO (World Trade Organization). 1994. *The Results of the Uruguay Round of Multilateral Trade Negotiations.* Geneva.

———. 1996. "Annexes to the Draft Protocol of China's Accession to the WTO." Geneva.

Zhang, Enshu. 1995. "The Reform of Foreign Exchange Administration System." *Great Economic Reforms Series.* Beijing: Zhongguo Jihua Chubanshe.

Zhang, Shu-guang, Yan-Sheng Zhang, and Zhong-xin Wan. 1996. "Measuring the Cost of Protection in China." Unirule Institute of Economics, Beijing.

chapter three

Integration with Global Capital Markets

China's sizable and growing financial integration with global capital markets is reflected in its large shares in cross-border flows to developing countries: 40 percent of foreign direct investment (FDI) flows, 10 percent of commercial bank lending, and rising portfolio equity and bond inflows.

China will need to attract sustained levels of FDI inflows and a more balanced composition of flows to achieve some of its important development objectives: faster development of its physical infrastructure, more efficient and better technologies, and higher flows of investment to interior and poorer provinces. To sustain FDI flows, China will need a stable macroeconomic environment and complementary reforms to improve the regulations and institutions governing FDI. To shift the composition of these flows toward its development priorities, China will need to improve the transparency of its investment regime, which investors view as one of the most complex in the world. Greater transparency in the rules governing FDI and better property rights for investors will attract a wider range of FDI and ensure greater competition. This change would in turn help ease the skewed geographical and sectoral distribution of FDI inflows, bring in and disseminate better technology, and attract more investment in longer-gestation infrastructure sectors. Trade policy reforms would be equally important to bring FDI into more efficient areas and to transfer greater benefits to the economy.

Pressures in domestic financial markets to integrate with international capital markets will also grow over time. Such integration will bring about greater efficiency in domestic financial intermediation. In that context, financial sector reform of both domestic bank and non-bank financial institutions will be important to integrating more efficiently with global financial markets. And reforms in capital markets will be essential to improve their efficiency and stability. China should move gradually, however, to integrate with global capital markets and to establish full convertibility of its capital account. Experience shows that private capital flows can be volatile and test governments' macroeconomic management. Weak financial and capital markets magnify these risks. Measures to improve commercial borrowing and debt management would also be prudent.

Global financial flows to developing countries

The early 1990s saw a dramatic surge in private capital flows to developing countries. Although there was a temporary slowdown in 1994—a result of the peso crisis in Mexico and the rise in interest rates in industrial countries—private capital flows have grown rapidly since (figure 3.1). In the previous peak year of 1981 (before the debt crisis) private flows were about 30 percent of the current level (or about 50 percent once adjusted for inflation). Moreover, current flows are equivalent to about 14 percent of fixed investment in recipient countries, double the rate before the debt crisis (and about three times higher than in 1990). There has also been a dramatic shift in composition, from bank lending to FDI and portfolio flows.

Driving factors

Long-term prospects for private flows remain bright. There is likely to be more globalization of production

(and hence FDI flows) because of greater competition and contestability of product markets worldwide. Greater competition and contestability are being driven, in turn, by falling trade and investment barriers and falling transport and communications costs. Diversification of institutional portfolios and financial innovation are also leading to increased private flows from developed to emerging markets. Rapid recovery after the Mexico crisis suggests that investor interest in emerging markets will continue. Moreover, with real interest rates in industrial countries likely to remain moderate, flows should continue. And growth in world trade and global production will promote continued large private flows—especially FDI—to developing countries.

Most private capital flows have gone to countries where progress on macroeconomic stabilization and structural and financial sector reforms has been greatest. Institutional investors are leading the way in portfolio flows. Global portfolio investment in emerging markets may have reached $170 billion in 1995, with U.S. institutional investors accounting for almost half.

Benefits

Increased private capital flows are associated with higher growth in recipient countries (figure 3.2). FDI can provide an engine for growth through its contribu-

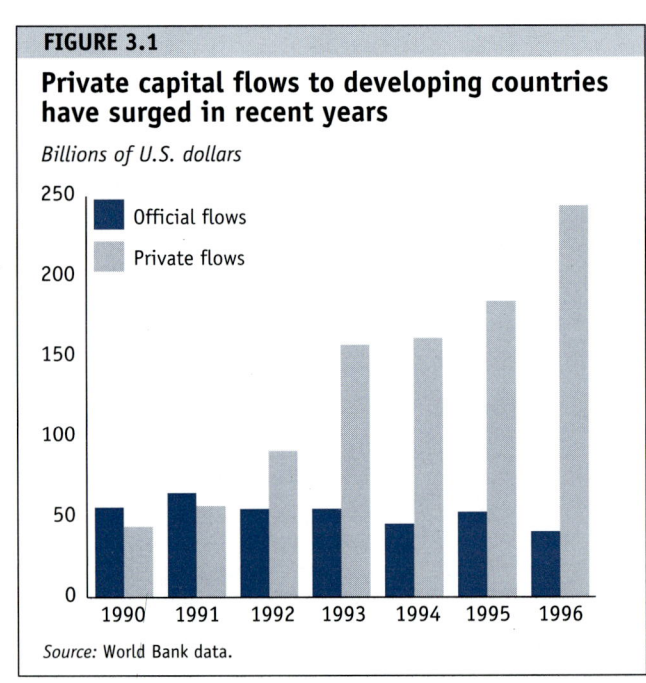

FIGURE 3.1

Private capital flows to developing countries have surged in recent years

Billions of U.S. dollars

Source: World Bank data.

tion to physical capital formation and, more important, to human capital development, transfer of technology and know-how, and expansion of markets and foreign trade. Greater portfolio investments and capital market integration generate benefits by lowering the costs of capital and increasing the efficiency of resource allocation. Causality can also work the other way, because faster growth attracts more capital.

The spillover effects (or externalities) of FDI on the host country work better in more open than in closed economies because of greater competition and fewer distortions. Opening up stock markets to foreign investment can help increase their efficiency. Competition from foreign financial institutions promotes the import of more sophisticated financial techniques, the adaptation of those techniques to the local environment, and greater investment in improving information processing and financial services.

Financial flows to China

China's financial integration with the world economy—measured by country risk ratings and by private capital flows and their diversification—has improved markedly since the mid-1980s. China now accounts for about 40 percent of FDI flows to developing countries and 10

percent of cross-border commercial debt flows. It also has begun to tap into growing portfolio investment flows, attracting 10 percent of international equity flows and 5 percent of international bond flows to developing countries. Even with the build-up of large foreign exchange reserves and high domestic savings, sustained external flows will be important to support China's development objectives. Prospects for further integration with world capital markets depend on major policy improvements, however.

China has seen a more than fivefold increase in capital inflows in the 1990s, led by FDI (figure 3.3). At $38 billion in 1995 (and an estimated $42 billion in 1996), FDI accounted for 25 percent of domestic investment, 13 percent of industrial output, 31 percent of exports, 11 percent of tax revenues, and 16 million jobs (table 3.1). The effects can be even greater at the provincial level. In Fujian Province, for example, foreign enterprises contributed more than 20 percent of GDP and employment and 58 percent of exports in 1996.

But incentive policies favoring joint ventures have fostered inefficiencies such as roundtripping (some 20 percent of FDI according to some estimates; see chapter 1). High protection also has attracted inefficient (in terms of scale and costs) investments, such as in automobiles. Moreover, FDI flows to China have been

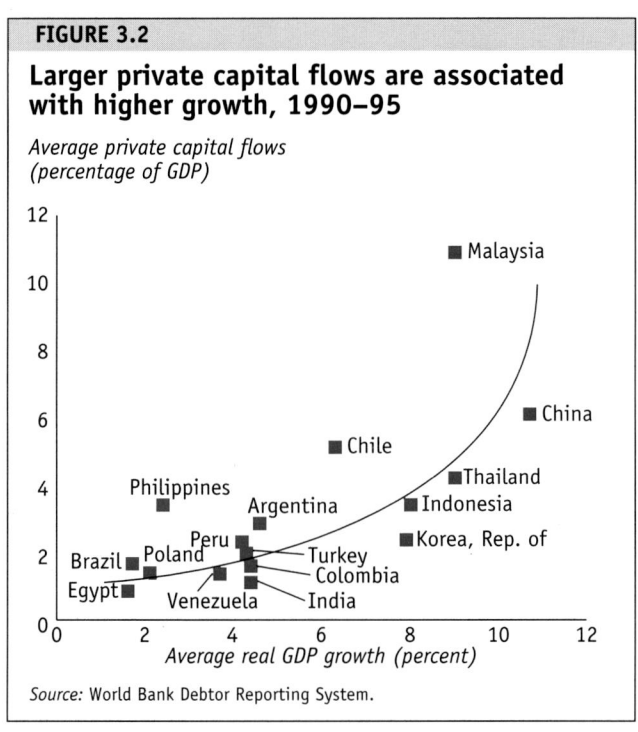

FIGURE 3.2

Larger private capital flows are associated with higher growth, 1990–95

Average private capital flows (percentage of GDP)

Source: World Bank Debtor Reporting System.

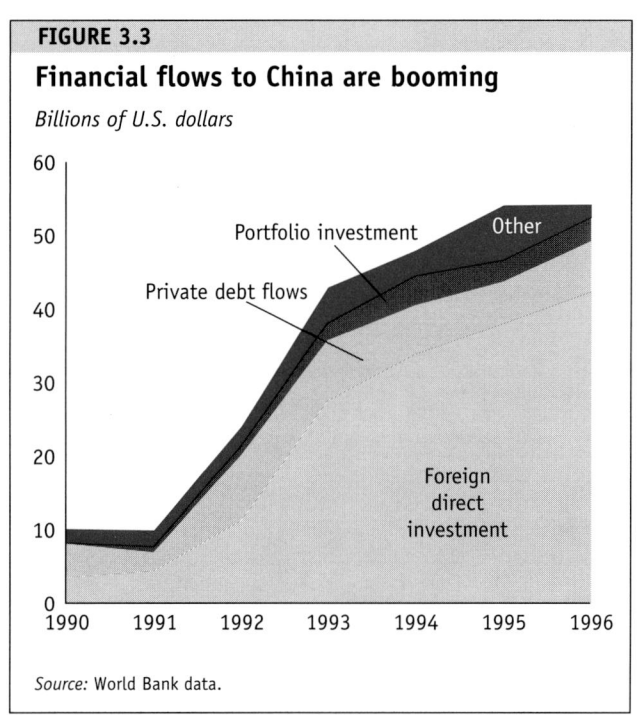

FIGURE 3.3

Financial flows to China are booming

Billions of U.S. dollars

Source: World Bank data.

TABLE 3.1
Foreign direct investment in China, 1991–95

Item	1991	1992	1993	1994	1995
FDI flows (billions of U.S. dollars)	4.4	11.2	27.5	33.8	37.5
Average amount per project (millions of U.S. dollars)	0.9	1.2	1.3	1.8	2.5
FDI as a share of gross domestic investment (percent)	4.5	8.0	13.6	18.3	25.0
Volume of exports by foreign affiliates (billions of U.S. dollars)	12.1	17.4	25.2	34.7	46.6
Share of exports by foreign affiliates in total exports (percent)	17	20.4	27.5	28.7	31.3
Share of industrial output by foreign affiliates in total industrial output (percent)	5	6	9	11	13
Number of employees in FDI projects (millions)	4.8	6	10	14	16
Tax contribution as share of total taxes (percent)	..	4.1	..	..	11.2

Source: China Ministry of Foreign Trade and Economic Cooperation, State Tax Administration; UNCTAD 1996; Word Bank Data Reporting System.

skewed toward coastal provinces. Twelve coastal provinces and municipalities—notably, Guangdong, Shanghai, Fujian, and Jiangsu—have attracted more than 90 percent of all flows. The sectoral distribution also has been biased, dominated by processing of tradable manufactures (and real estate), with few links to the rest of the domestic economy. Finally, the bulk of FDI inflows come from a narrow source of mainly overseas Chinese-owned firms, with Hong Kong, China, accounting for 60 percent of recent inflows.

Cross-country evidence suggests that the most important determinant of FDI flows is a stable macroeconomic environment, complemented by strong growth performance. To ensure continues flows, China will also need to improve the regulatory and institutional framework for FDI and increase the transparency of tax and foreign exchange rules. Special fiscal incentives to attract FDI generally are unlikely to succeed and would be costly. More important to multinational corporations are market size, economic growth, production costs, skill levels, infrastructure development, political and economic stability, and the regulatory framework. China is moving toward leveling the playing field for domestic and foreign firms. Recent efforts to unify the tax system, eliminate import duty exemptions for foreign affiliates, and restrict tax incentives at the province and city levels are all steps in the right direction.

China also needs a better policy framework for FDI—one that encourages firms to bring in best practices and know-how from world markets and disseminate them throughout the economy. This means greater transparency in the investment regime, more open trade policies, and better property rights for investors.

China has tried to attract greater technological benefits by specifying formal technology transfer requirements (such as use of the most advanced technology available, local research and development, and access to patents or transfer of skills to local staff and firms). In practice, however, such requirements make little difference. The best way to encourage foreign firms to bring in advanced technologies is to maintain open and competitive markets, because competition forces firms to be more efficient (box 3.1). More open trade and investment regimes, which are crucial factors in attracting more foreign firms to invest, are also the best channels for encouraging such competition. It is also vital to encourage competition from all sources and to eliminate any discrimination by type of foreign investors.

China needs to diversify its FDI sources. So far FDI inflows to China have been dominated by investment flows from overseas Chinese-owned firms. Diversification will help make FDI flows more sustainable and bring wider benefits (for example, expanding global production and marketing links to the source economies of FDI).

Another challenge for China is to induce greater use of FDI in infrastructure development. This would bring in better management and improve service standards in critical areas. Given the size and complexity of the proposed investment program—$600 billion over the next decade—traditional public supply and financing of infrastructure will be inadequate. Although some infrastructure projects have been promising, they have attracted only limited FDI. There are three reasons investors have shied away:

• *Enabling environment, and legal and regulatory restrictions.* Fewer projects have been implemented

Open policies lead to technological and productivity benefits from foreign investment

More open trade policies are associated with the presence of foreign firms (proxied by FDI to GDP ratios) and economywide technological and productivity gains in developing countries. There is also a strong positive correlation between FDI to GDP ratios and diversification in high-technology exports in countries with open trade regimes. Finally, there is a strong positive correlation between FDI to GDP ratios and measures of total factor productivity growth in countries with open trade policies.

Source: Dasgupta and Imai 1997.

Correlation between FDI and diversification in high-tech exports, 1970–95

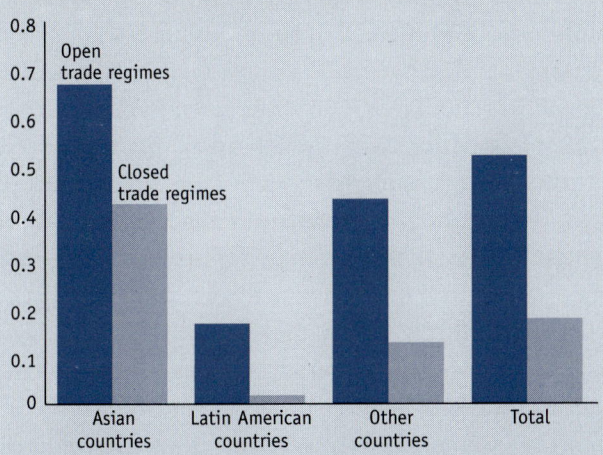

Correlation between FDI and total factor productivity growth, 1970–95

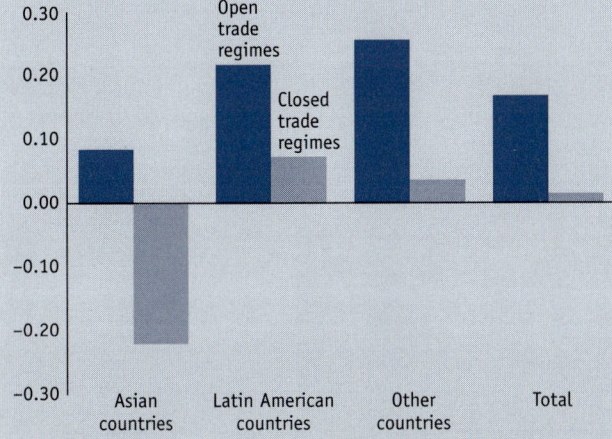

Note: Sample for the top figure covers fifty countries. Sample for the bottom figure cover sixty-nine countries.
Source: IMF, *International Financial Statistics;* Sachs and Warner 1995; Coe, Helpman, and Hoffmaister 1994; World Bank data and staff estimates.

than might otherwise have attracted investors because of the weak institutional framework and the absence of enabling legislation. The perception of a cap on profitability also has hurt infrastructure proposals in China. Experience elsewhere suggests that governments should use competitive bidding that gives priority to the cost of service or specifies an upper price (on power sales or road tolls, for instance).

• *Procedures and policy uncertainties.* National and provincial decisionmaking and the division of responsibilities for approving FDI projects have delayed foreign investment projects. Establishing a coherent national strategy on FDI in infrastructure and clear decisionmaking rules will be crucial.

• *Performance by state-owned enterprises.* Because full foreign ownership is not encouraged, FDI projects typically involve state agencies. Performance obligations have been difficult to obtain from these agencies. Another problem is that project performance guarantees are not granted under Chinese policy, making commercial lending difficult. Improvements are also needed in laws and regulations relating to mortgages, loan security, contract enforcement, and foreign exchange access.

Many of these issues are being gradually resolved, and the number of foreign investment projects in power generation, ports, highways, and railways is edging upward. The Philippines is a good example of how improvements in the policy framework can lead to considerable investor interest in infrastructure.

Global integration of capital markets

Portfolio capital flows to China have been limited, but between 1992 and 1994 they climbed from negligible levels to about 0.8 percent of GDP. There was a drop in 1995 following the Mexico peso crisis, but flows revived quickly in 1996. Indirect investments in China's capital markets—through, for example, investment in companies listed on Hong Kong's (China) stock market with substantial Chinese assets—have been growing rapidly. There are likely to be greater pressures for further integration of China's capital markets with the rest of the world. Such integration would improve the efficiency of financial intermediation. Thus the key policy issues are how the authorities should respond to these pressures for faster integration and how fast China should open up its capital account.

The opening of exchanges in Shanghai in 1990 and in Shenzhen in 1991 marked the beginning of a securities market in China. To attract foreign portfolio investors, a separate and limited class of B shares open only to foreign investors was introduced in 1992. B shares have attracted about $2 billion in portfolio investment in seventy-two stocks listed on local exchanges. Although foreign participation in B shares helped add liquidity, the large price discrepancy between A shares (which foreigners cannot own and which command a high premium) and B shares undermines the efficiency of China's capital markets. The price differential between the two also implies that there is a large amount of capital inside China seeking a home. Access of foreign investors to Chinese equities has been enhanced with the introduction of H shares on the Hong Kong (China) Stock Exchange for companies with large Chinese holdings. The authorities have also permitted selected Chinese companies to issue American depository receipts (ADRs) and global depository receipts (GDRs) to tap U.S. and other equity markets.

Unlike stock markets, China's fixed-income securities markets remain tightly closed to foreign investors. And it is unlikely that foreign investment will be permitted in yuan-denominated debt instruments in the medium term. But China's annual volume of new overseas bond issues has grown fast, soaring from $130 mil-

lion in 1990 to almost $4 billion in 1994. After a slowdown in 1995 following the Mexico peso crisis, international bond issues have picked up strongly (figure 3.4). China has been rated investment grade by all the major rating agencies. Still, China's securities market, one of the fastest growing in the 1990s, suffers from serious distortions in pricing and market development. Share and bond prices are volatile. And relative to GDP, the development of China's capital markets remains far behind that of other East Asian economies (figure 3.5).

The internationalization of China's capital markets needs to proceed gradually and in tandem with financial development. This process will take time. Banking and financial sector reforms should include government withdrawal from direct interventions in credit markets, elimination of interest rate controls, transformation of banks into commercial institutions, entry of private banking and financial institutions, and adequate accounting standards, corporate governance, and prudential regulations. In the nonbank financial sector reforms need to include modern laws on institutions, investor protection, information disclosure, and prudential and regulatory oversight. Further development of domestic capital markets will come with increases in efficiency, stability, and transparency. Different classes of equity need to be harmonized. Greater flexibility in interest rates will help the orderly issue of fixed-income securities. Standards can be set for domestic credit rat-

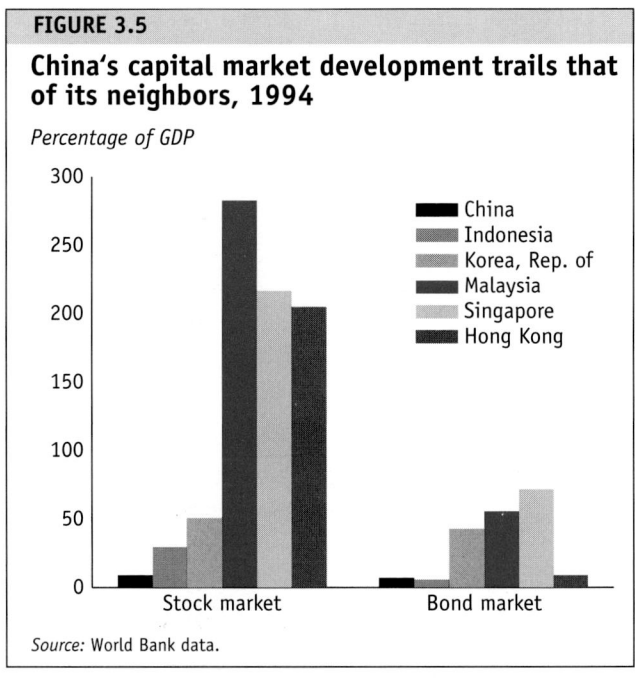

FIGURE 3.4

China's international bond issues have soared since the early 1990s

Millions of U.S. dollars

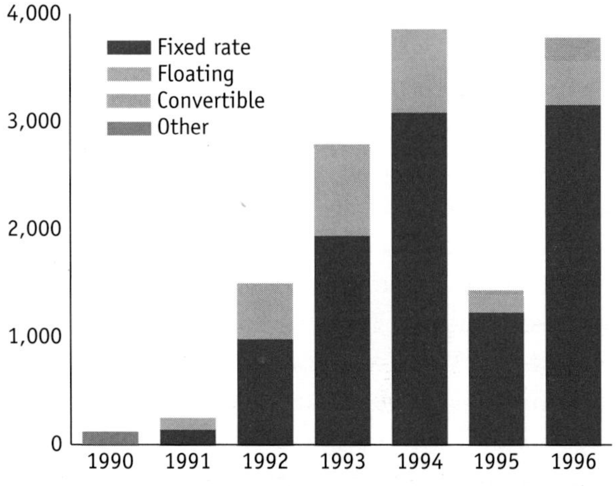

Legend: Fixed rate, Floating, Convertible, Other

Source: World Bank data.

FIGURE 3.5

China's capital market development trails that of its neighbors, 1994

Percentage of GDP

Legend: China, Indonesia, Korea, Rep. of, Malaysia, Singapore, Hong Kong

Source: World Bank data.

ing agencies, competition should be encouraged among underwriters, and the regulatory regime and oversight should be strengthened. Stock exchanges alone should enforce eligibility criteria for listing.

The convertibility of the yuan on the capital account should also come with the opening of domestic markets and integration with world markets, but China needs to approach financial liberalization with caution (box 3.2). International experience shows that capital flows can be volatile and can test a government's macroeconomic management. More important, a weak financial sector magnifies the risks associated with capital flows, since banks may tend to get involved in riskier projects or undertake complex financial deals (such as in derivatives).

Commercial borrowing and debt management

China's improved creditworthiness has made it the main beneficiary of syndicated lending to developing countries, accounting for about 10 percent of cross-border commercial debt flows in 1996. Despite commercial banks' continued cautious attitude toward emerging markets, international commercial loans to China rose from $2.9 billion in 1990 to $12 billion in 1996, thanks to increased project financing for infrastructure development.

Even with the rapid build-up of external debt (to around $130 billion at the end of 1996), strong macroeconomic performance makes China's debt indicators (around 85 percent for debt to exports and 20 percent for debt to GNP) very creditworthy. China's external debt indicators stand at less than half the developing country average and are among the lowest in the region. Roughly 58 percent of external debt is denominated in U.S. dollars, 21 percent in Japanese yen.

Within this comfortable and conservative borrowing framework, however, some important reforms are possible. First, enterprise reforms will soon make the foreign borrowing plan redundant. But if state enterprises are to borrow abroad on their own account, they should be required to meet certain criteria. These should include a modern corporate structure, use of modern accounting systems, audit of accounts according to international standards, profitable operations for the past three years or so, and a demonstrated ability to service debt.

With regard to international bond issues, reforms to improve the current "window" system could be important. China's international bonds, which amounted to about $4 billion in 1996 (5 percent of the developing country total), are issued through selected banks and trust and investment corporations. These institutions act as intermediaries, onlending to domestic users (although some borrowing is for their own use). Three international trust and investment companies—including CITIC—are the most active, with strong balance sheets accounting for some 30 percent of China's external bond issues (and more than half of nongovernment bond issues). The procedures followed at these "windows" do not necessarily ensure efficiency in the use of such borrowing. Proceeds from foreign bond issues may be administratively allocated, and onlending terms are not clearly defined. Overseas borrowing is supposed to reduce costs, but the system has some disadvantages—weak balance sheets of some intermediaries and, for those who are excluded from windows, encouragement of offshore borrowing through foreign subsidiaries. Since 1993 the authorities have been considering allowing domestic enterprises direct access to overseas bond markets (provided they meet certain well-defined criteria) and allowing international credit rating agencies a bigger role in assessing the financial adequacy of borrowers.

There are also some issues that need careful consideration in the management of China's external debt. One is the monitoring of external debt through timely and accurate data collection and analysis by the People's Bank. A clear framework needs to be developed for nonrecourse borrowing from foreign institutions to finance infrastructure. (The World Bank's expanded guarantee program could be useful.) And there is a need to integrate external debt management more closely with a public finance framework that considers the total financing requirements (domestic and external) of the public sector.

China's outward investments

Between 1989 and 1995 China's officially approved direct and portfolio investment abroad totaled $17 billion. In 1995 China's outward investment accounted for about 2 percent of global capital flows, making it the eighth largest supplier among all countries and the largest outward investor among developing countries. But China's capital outflow is probably much higher than the official figure. The largest outward investments are in Hong Kong (China).

Several factors account for China's rising outward investments:

• *Macroeconomic.* China's domestic savings have exceeded domestic investment needs, as reflected in a current account surplus of more than $20 billion in 1996. Large capital outflows are an outcome of high domestic savings, large foreign capital inflows, and high official reserves.

• *Benefits of overseas investment.* Overseas investments offer higher returns and other benefits, including better access to markets (more than half of overseas investments are trade related), natural resources (mining, forestry investments), and technology (manufacturing investments in the United States and Canada). And foreign assets are a means of portfolio diversification—of reducing risk and improving returns to assets.

• *Economic relationships.* China's overseas investments also helped strengthen its relationships with other economies, such as those in Africa, and, most important, Hong Kong leading up to its reunification.

• *Policy distortions.* Chinese enterprises established affiliates in Hong Kong so that they could invest in China as foreign firms, thereby taking advantage of tax and other incentives.

A potentially serious problem is the lack of adequate monitoring of outward investments. State funds have sometimes been wasted on inappropriate projects: one-third of Chinese investments abroad are loss-making, according to some reports. New regulations are being drafted to provide a general framework in which future overseas investment will be approved. There is an urgent need for procedures that establish clear rules, especially transparency in control, accountability, and information; improve ownership structures, governance, and management of state-owned enterprises; set regulatory or screening mechanisms to protect against large risks; and eliminate distortions in taxes, exchange rates, and credit markets that lead to roundtripping. But there is a need to avoid centrally dictated investments, as they are unlikely to solve the performance problem and could distort investments further.

References

Aitken, Brian, and Ann Harrison. 1993. "Does Proximity to Foreign Firms Induce Technology Spillovers?" World Bank, Washington, D.C.

Blomström, M., R. Lipsey, and M. Zejan. 1992. "What Explains Developing Country Growth?" NBER Working Paper 4132. National Bureau of Economic Research, Cambridge, Mass.

———. 1996. "Is Fixed Investment the Key to Economic Growth?" *Quarterly Journal of Economics* (February).

Borensztien, E., J. DeGregorio, and J. Lee. 1994. "How Does Foreign Direct Investment Affect Economic Growth?" IMF Working Paper WP/94/110. International Monetary Fund, Washington, D.C.

Broadman, Harry, and Xiaolun Sun. 1996. "The Distribution of Foreign Direct Investment in China." World Bank, Washington, D.C.

Buckberg, Elaine. 1993. "Emerging Stock Markets and International Asset Pricing." In Stijn Claessens and Sudarshan Gooptu, eds., *Portfolio Investment in Developing Countries.* World Bank Discussion Paper 228. Washington, D.C.

Coe, David, Elhanan Helpman, and Alexander Hoffmaister. 1994. "North-South R&D Spillovers." IMF Working Paper WP/94/144. International Monetary Fund, Washington, D.C.

De Gregoiro, José, and Pablo Guidotti. 1995. "Financial Development and Economic Growth." *World Development* 23 (March): 433–48.

Fry, Maxwell. 1993. "Foreign Direct Investment in a Macroeconomic Framework: Finance, Efficiency, Incentives, and Distortions." Policy Research Working Paper 1141. World Bank, Washington, D.C.

Graham, Edward M. 1996. *Global Corporations and National Governments.* Washington, D.C.: Institute for International Economics.

Husain, Ishrat, and Kwang W. Jun. 1992. "Capital Flows to South Asia and ASEAN Countries: Trends, Determinants, and Policy Implications." Policy Research Working Paper 842. World Bank, Washington, D.C.

IMF (International Monetary Fund). 1995. "Capital Account Convertibility." Occasional Paper 131. Washington, D.C.

Jun, Kwang W. 1993. "Effects of Capital Market Liberalization in Korea: Empirical Evidence and Policy Implications." In Stijn Claessens and Sudarshan Gooptu, eds., *Portfolio Investment in Developing Countries.* World Bank Discussion Paper 228. Washington, D.C.

Jun, Kwang W., and Amitava Sardar. 1996. "Internationalization of Emerging Capital Markets: Cross-country Experience." World Bank, International Economics Department, Washington, D.C.

Kim, E. Han, and Vijay Singal. 1993. "Opening up of Stock Markets by Emerging Economies: Effect on Portfolio Flows and Volatility of Stock Prices." In Stijn Claessens and Sudarshan Gooptu, eds., *Portfolio Investment in Developing Countries.* World Bank Discussion Paper 228. Washington, D.C.

Levine, Ross, and Sara Zervos. 1995. "Stock Markets, Banks, and Economic Growth." World Bank, Washington, D.C.

Lipsey, R., M. Blomström, and I. Krauis. 1990 "R&D by Multinational Firms and Host Country Exports." In Robert E. Evanson and Gustav Ranis, eds., *Science and Technology: Lessons for Development Policy.* Boulder, Colo.: Westview Press.

Sachs, Jeffrey, and Andrew Warner. 1995. "Economic Convergence and Economic Policies." NBER Working Paper 5039. National Bureau of Economic Research, Cambridge, Mass.

UNCTAD (United Nations Conference on Trade and Development). 1996. *World Investment Report 1996.* Geneva.

Wall, David. 1996. "Outflows of Capital from China." OECD Development Center Technical Paper. Organisation for Economic Co-operation and Development, Paris.

Wei, Shang-Jin. Forthcoming. "Foreign Direct Investment in China: Sources and Consequences." In Takatoshi Ito and Anne Krueger, eds., *Financial Deregulation and Integration in East Asia.* University of Chicago Press.

World Bank. Various years. *Global Development Finance.* (formerly *World Debt Tables*) Washington, D.C.

———. Various years. *Global Economic Prospects and the Developing Countries.* Washington, D.C.

chapter four

Global Effects of China's Integration to 2020

ver the past twenty-five years China's GDP growth averaged 10 percent a year and its share of world trade tripled to about 3 percent. While its pace of economic growth is expected to slow to more sustainable levels, China's role in the world economy in the next quarter century should increase further. Its share in world trade will more than triple by 2020 (to about 10 percent), and it is expected to become the second largest trading nation (after the United States). China will account for some 40 percent of the increase in all developing country imports between 1992 and 2020 and serve as an engine of growth for world trade. According to our projections, the implications of China's growing trade and integration for the rest of the world are enormous:

• Industrial countries will benefit from China's growth because of China's rising demand for imports of capital- and knowledge-intensive manufactures and services and primary products and because of significant terms of trade gains as a result of its rising demand for such products.

- For neighboring developing countries that are ahead of China and not close competitors (Asia's newly industrialized countries, such as the Republic of Korea), there are likely to be significant gains all around.
- For countries that are close competitors of China (low- and middle-income Asian countries such as India, Indonesia, the Philippines, and Thailand), world market shares and volumes of trade will likely continue to expand, but significant terms of trade losses are expected in their main exports of labor-intensive manufactures. Net income gains, however, will still be large.
- For developing countries that do not directly trade much or compete with China (those in Latin America, Sub-Saharan Africa, and Europe and Central Asia), there will be neither significant gains nor major losses.
- If it were to integrate more slowly, the greatest adverse effects would be on China itself, but there would also be significant welfare losses for the rest of the world.

Evaluating how China's growing trade will affect the world

There are three tools for evaluating the effects of China's growing trade integration on other countries: simple trade theory to inform us of the likely directions of change, an empirical model of multiregional trade, and the historical record. Although all three approaches are used, this chapter relies primarily on the Global Trade Analysis Project (GTAP) database and model projections of trade patterns for 2020, compared with the base year 1992.[1] The assumptions underlying the model are set out in box 4.1.

The projections reflect many assumptions, and the increasing importance of China is only one part of a complex picture. It is nevertheless instructive to review China's role as it might evolve under various scenarios. Applying the economic growth, trade policy liberalization, and transport cost reduction assumptions set out in box 4.1 yields a projection of world trade volume growth under the baseline scenario of about 5.5 percent a year in 1992–2020 (similar to that since the mid-

---

BOX 4.1

Growth assumptions about the world in 2020

To project the likely structure of world trade (and the changes in terms of trade) in 2020, several assumptions are made about the primary sources of growth in the world's economies (World Bank 1997).

- Industrially mature economies are expected to grow 2.1–2.6 percent a year, or at about the current long-term rates of growth, as increases in human capital make up for the slowdown in growth of the labor force. Dynamic newly industrialized economies in Asia (Hong Kong, China, Korea, Singapore, and Taiwan, China) are assumed to continue to grow vigorously, albeit at a slower pace of about 4.5 percent a year. These economies are unlikely to slow down to industrial country growth rates because of continued high savings rates and scope for capital deepening. High-income economies are expected to grow about 2.5 percent a year.
- Developing countries are expected to grow about 5.4 percent a year, twice as fast as high-income economies. China is expected to grow 7 percent a year, reflecting an easing of total factor productivity growth to 2–3 percent from the 4–5 percent during 1984–94. Other East Asian developing countries are expected to continue their current vigorous growth rates of slightly over 7 percent a year for the next twenty-five years. For the rest of the world, current expectations about long-term growth in World

Bank forecasts up to 2005 are used and reflect long-term growth potential of about 4.2–5.5 percent a year.
- Baseline projections go beyond the Uruguay Round tariff reductions and abolition of the Multi-Fibre Arrangement (by 2005) by assuming that developing country tariffs for manufactures will average those of OECD countries by 2020. China's accession to the World Trade Organization is also incorporated (on the basis of China's April 1997 offers).
- Services sector productivity gains are expected to be slower than for manufactures in all countries and productivity in agriculture is assumed to rise faster than in nonagriculture. Global transport costs are assumed to fall 2 percent a year. Real export prices for energy and agricultural products are derived from special studies undertaken by the World Bank.
- Under these growth assumptions China's share of world GDP (measured at 1992 prices and market exchange rates) rises from 1 percent in 1992 to nearly 4 percent in 2020. In purchasing power parity (PPP) exchange rate terms (but adjusting for likely changes in PPP exchange rates as China becomes richer), China's share of world GDP rises from 4 percent in 1992 to 8 percent in 2020, when China becomes the second biggest economy in the world after the United States (whose share of world GDP is estimated at 19 percent).

TABLE 4.1
Trade growth and market shares, 1992–2020
(percent)

Region	Exports			Imports		
	Growth rate, 1992–2020	Share of world 1992	Share of world 2020	Growth rate, 1992–2020	Share of world 1992	Share of world 2020
World	5.5	100.0	100.0	5.3	100.0	100.0
High-income	4.0	76.5	51.6	4.3	74.3	56.6
OECD	3.5	67.8	40.4	4.0	65.3	45.3
Newly industrialized economies	6.5	7.4	9.7	6.3	7.2	9.4
Hong Kong (China)	6.0	1.3	1.5	5.7	1.8	1.9
Low- and middle-income	8.1	23.5	48.4	7.3	25.7	43.4
China	10.0	3.0	9.8	10.2	2.8	9.9
Brazil	7.2	1.2	1.9	6.8	0.9	1.3
India	12.0	0.7	3.9	11.0	0.8	3.2
Indonesia	8.8	1.1	2.7	7.8	0.9	1.8
Transition economies	6.2	3.0	3.6	5.9	3.4	3.9
ASEAN3[a]	9.6	2.8	8.4	8.6	3.0	7.0
Rest of South Asia	8.0	0.5	0.9	6.8	0.6	0.8
Rest of Latin America	6.7	2.8	3.9	5.4	3.5	3.5
Sub-Saharan Africa	6.7	1.7	2.4	5.3	2.1	2.1
Middle East and North Africa	6.4	5.2	6.6	5.4	5.9	6.0
Rest of world	9.4	1.5	4.2	8.1	1.9	3.9

Note: Exports and imports are in constant 1992 dollars.
a. Malaysia, Philippines, and Thailand.
Source: World Bank 1997.

1980s). The high-income OECD countries' share of world exports and imports falls from 65–70 percent in 1992 to 40–45 percent in 2020, while that of developing countries doubles to 45–50 percent (table 4.1). China's share of global trade rises from 3 percent to about 10 percent, and China becomes the second largest trading country in the world, only slightly behind the United States (12 percent of world exports) but much ahead of Japan (5 percent of world exports). Another measure of China's importance is that it contributes about 40 percent to the expected rise in total world imports by developing countries between 1992 and 2020, serving as an engine of world trade.

Industrial countries

Trade theory suggests that the effects on industrial countries of trade with developing countries, such as China, are mostly favorable, except in special circumstances.[2] China's rapid growth and integration with the global economy are expected to lead to faster growth in import demand for capital- and knowledge-intensive products and services from industrial countries. This rising demand may be accompanied by favorable terms of trade effects (as the prices of capital- and knowledge-intensive products rise relative to prices of labor-intensive products). There can be some adverse effects on industrial country employment and wages, however, if growing imports of labor-intensive manufactures displace unskilled labor and depress relative wages.

The Global Trade Analysis Project database and model projections of trade patterns for 2020, when compared with the base year 1992, broadly support these conclusions:

• Industrial country (North America, Western Europe, and Japan combined) exports to China will grow about 6.5 percent a year. This increase will accelerate their overall annual export growth from roughly 3.5 percent (excluding China) to 3.7 percent (including China). While most of the export volume gains come from capital- and knowledge-intensive products and services, a significant part originates with primary products (such as food grains) in which some countries (especially in North America) have a comparative advantage.

• There are also significant terms of trade gains, as the prices of industrial country exports improve relative to the prices of labor-intensive manufactures imports. The cumulative gain between 1992 and 2020 is largest for

Japan (14 percent), followed by the European Union (12 percent) and North America (8 percent).

• Industrial countries lose world market shares in labor-intensive manufactures (from about 56 percent in 1992 to about 25 percent by 2020). Because much of the adverse impact on unskilled labor had already occurred by 1992, and the labor force in industrial countries has become more skilled, the relative wages of unskilled workers will not deteriorate further. In fact, the model suggests a small improvement in relative wages of unskilled labor in industrial countries between 1992 and 2020 (figures 4.1 and 4.2).

The model's results of only small adverse effects of increased labor-intensive exports from China and other developing countries on unskilled wages in industrial countries are not surprising. But the issue has attracted attention because the rapid rise in the global integration of developing countries in the past twenty-five years has coincided with three problems in industrial countries: rising de-industrialization, unemployment, and wage inequalities. Coincidence does not, however, mean causation. Trade accounts for less than 0.5 percentage point of the 15 percent drop in wages of U.S. unskilled workers between 1973 and 1993; the rest is due to technical changes that are biased against unskilled labor (Krugman 1994). Moreover, OECD trade with low-wage countries (China and other dynamic East Asian countries) is small, accounting for 2 percent of the GDP of OECD countries (OECD 1996a). Thus the observed effects were also small. Significant effects were confined to a few industries, such as textiles, clothing, footwear, and electronic goods.[3]

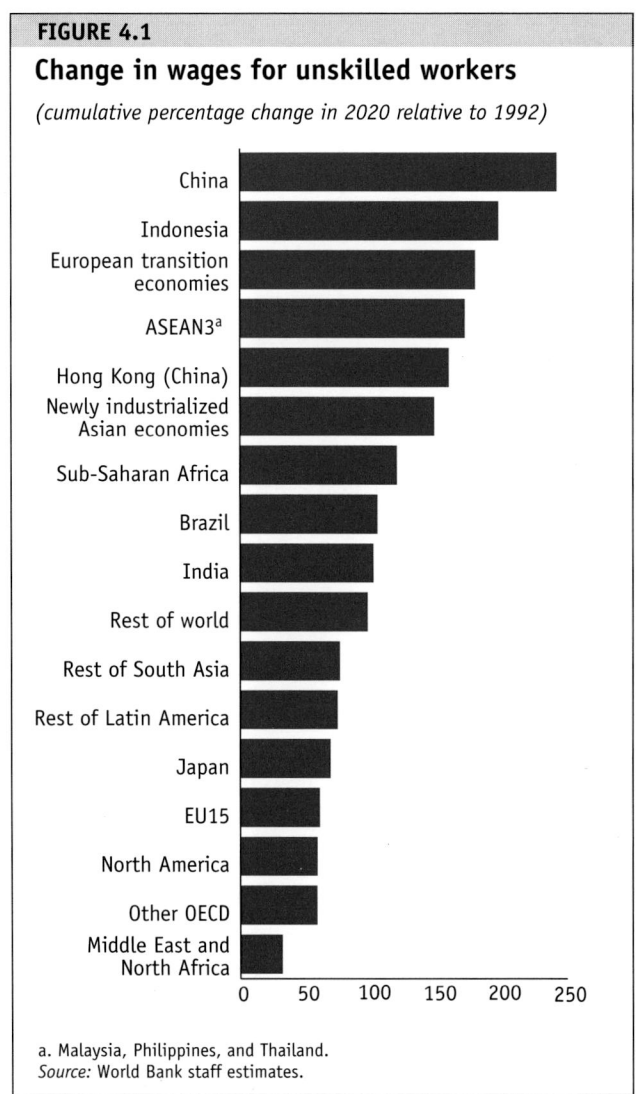

FIGURE 4.1

Change in wages for unskilled workers

(cumulative percentage change in 2020 relative to 1992)

a. Malaysia, Philippines, and Thailand.
Source: World Bank staff estimates.

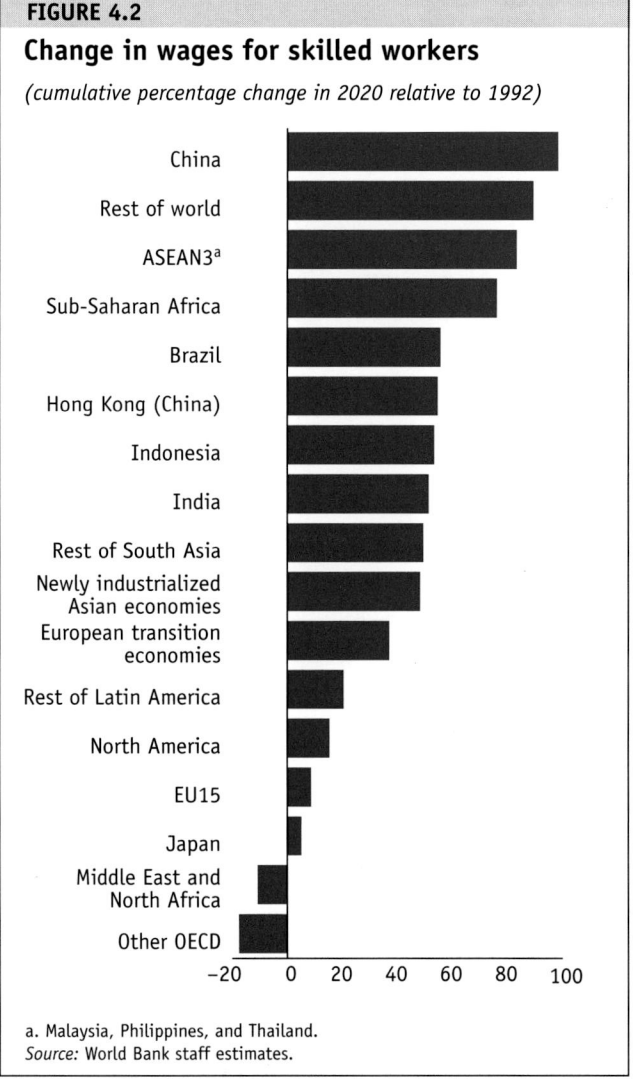

FIGURE 4.2

Change in wages for skilled workers

(cumulative percentage change in 2020 relative to 1992)

a. Malaysia, Philippines, and Thailand.
Source: World Bank staff estimates.

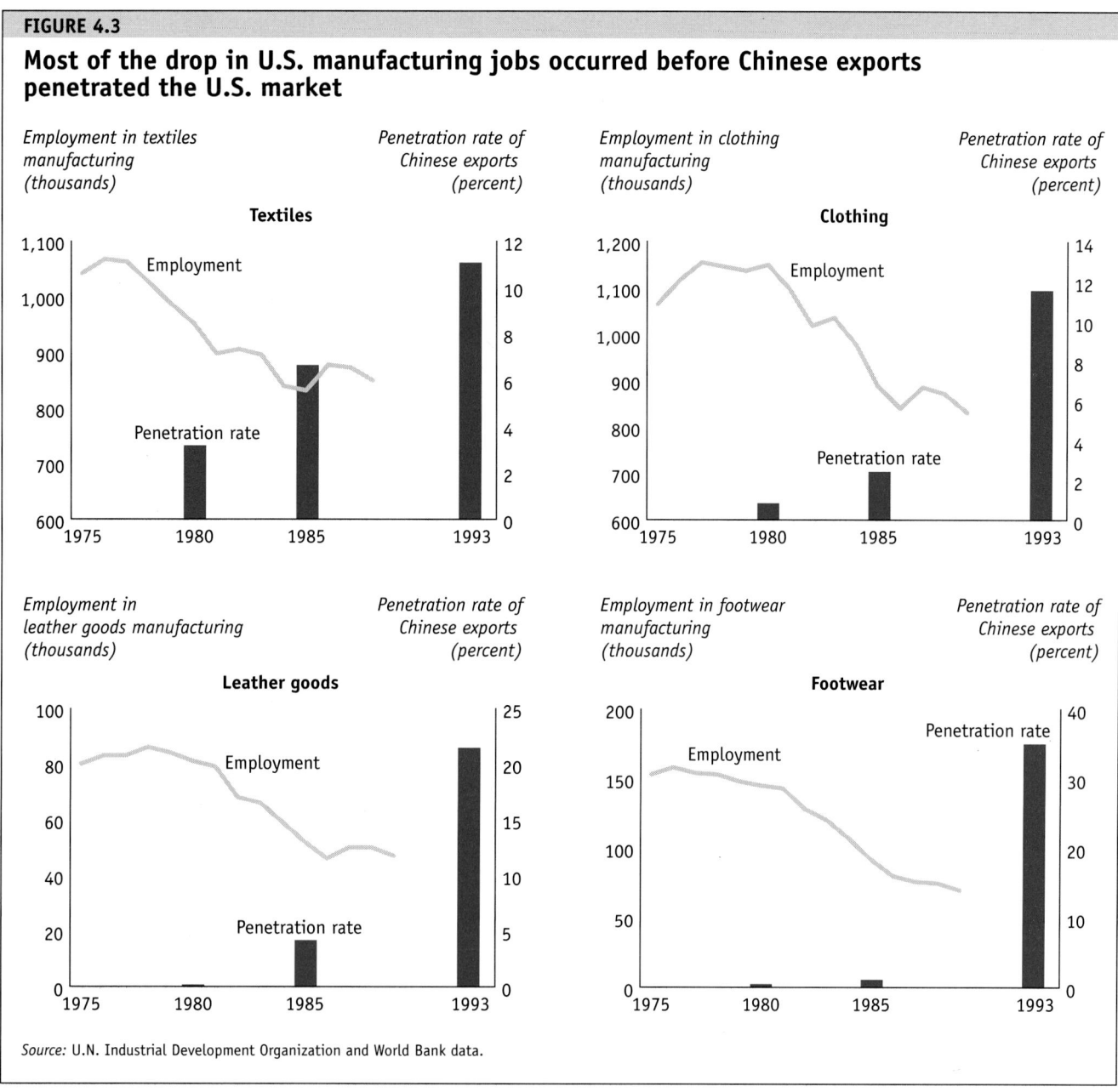

FIGURE 4.3

Most of the drop in U.S. manufacturing jobs occurred before Chinese exports penetrated the U.S. market

Employment in textiles manufacturing (thousands)

Penetration rate of Chinese exports (percent)

Textiles

Employment in clothing manufacturing (thousands)

Penetration rate of Chinese exports (percent)

Clothing

Employment in leather goods manufacturing (thousands)

Penetration rate of Chinese exports (percent)

Leather goods

Employment in footwear manufacturing (thousands)

Penetration rate of Chinese exports (percent)

Footwear

Source: U.N. Industrial Development Organization and World Bank data.

There does appear to be an association between rising penetration of Chinese exports to industrial countries and declining employment (figure 4.3). There have been three phases of growth in Chinese exports to industrial country markets, led by textiles in the early 1980s, followed by clothing in the mid-1980s and more recently by leather and footwear products. The fall in industrial country employment in each of these sectors began much earlier, however, as China simply replaced earlier penetration by Hong Kong (China), Korea, Singapore, and Taiwan (China), and as a structural decline in employment took place in labor-intensive manufacturing.

Developing countries

Greater Chinese trade integration should also be mostly favorable for other developing countries. A fast-growing China will import more goods that it cannot produce competitively, and countries able to supply its markets will gain. Strong competition by China in labor-intensive manufactures, however, could limit markets for other countries wanting to do the same. Even so, the results of alternative modeling approaches suggest that the world economy can absorb expansion of manufactured exports by all developing countries (Rodrigo and Martin 1996; Martin 1993).

Factor endowments and specialization. How will future trading patterns evolve? Among major developing countries, China is the most poorly endowed with land except for Korea (table 4.2). Thus, assuming that land supply remains constant, China's specialization in labor-intensive manufacturing relative to agriculture is expected to be the greatest. This will force China to import food and to move into manufacturing exports sooner to feed and create jobs for a growing population. Its high savings rate enables the build-up of capital needed to make the transition. At the same time, it will lose market shares in primary products.

What manufactured goods China exports will also depend on the quality of the labor force. With more people educated up to the secondary school level than to the tertiary level, and with low capital per worker, China is more likely to focus on low-skilled manufactures and light industry. Its closest competitors are India, Indonesia, the Philippines, and Thailand. With its weaker higher education base, China is unlikely to emerge as a major source of knowledge-based and complementary skilled labor products.

Confirmation through revealed comparative advantage. China's comparative advantage has been studied using indexes of revealed comparative advantage.[4] Comparing these indexes for 1986–87 and 1992–94 for China's major exports (accounting for about 80 percent of the total in 1992–94), China's advantage in toys, footwear, and travel goods remained more or less intact. But in clothing and telecommunications equipment other countries are closing in (India, the Philippines, and Indonesia).

Simulation results. Baseline projections suggest that China's comparative advantage will strengthen in intermediate technology manufacturing (tables 4.3 and 4.4). Together, mainland China and Hong Kong (China) are expected to lose 16 percent of the world market for apparel (and the industrial countries, 19 percent). This slack is picked up by India, Indonesia, the Philippines, and Thailand. More distant regions (Latin America and Sub-Saharan Africa) gain little. China also loses market shares across a range of primary agricultural products—grains, livestock, and processed food. This partly benefits industrial countries (those in North America), but also China's Asian neighbors (India, Indonesia, Malaysia, the Philippines, and Thailand) and Latin America.

China is set to move up the value-added ladder by gaining a 10 percent market share in light manufactures (such as leather, fabricated metal, and other products) and 8 percent in transport equipment and other machinery. In these sectors China's market share (and that of other lower-middle-income Asian countries) is expected to grow, displacing industrial countries and Korea and Taiwan (China). In capital-intensive heavy

TABLE 4.2

Factor endowments in China and selected other countries

Country	Cropland (hectares per worker) 1995	Labor force (millions) 1995	Labor force (millions) 2020	Physical capital (thousands of 1995 U.S. dollars per worker) 1995	Physical capital (thousands of 1995 U.S. dollars per worker) 2020	Tertiary education (years per capita) 1995	Tertiary education (years per capita) 2020
China	0.12	808.3	988.6	1.6	13.2	0.2	0.4
Indonesia	0.19	119.7	174.9	2.7	21.9	0.5	1.6
Korea, Rep. of	0.07	31.8	35.9	21.5	115.2	3.0	6.8
Malaysia	0.42	11.6	19.7	15.8	139.6	0.5	2.2
Philippines	0.20	40.1	72.1	3.0	27.2	2.7	5.0
Singapore	0.00	2.1	2.6	62.7	384.9	0.5	2.7
Thailand	0.58	39.9	53.9	7.9	73.2	1.3	3.1
Brazil	0.61	101.4	145.9	9.6	31.2	1.1	2.1
India	0.30	561.3	886.2	1.7	8.2	0.5	1.3
Japan	0.05	87.0	75.1	131.4	397.8	0.8	1.8
United States	1.09	172.3	200.2	73.0	—	2.1	3.5

Source: Ahuja and Filmer 1995; World Bank staff estimates.

TABLE 4.3
Revealed comparative advantage indexes, China and comparators, 1986–94
(index: China = 100)

Country	Clothing (841) 1986–87	Clothing (841) 1992–94	Telecommunications equipment (724) 1986–87	Telecommunications equipment (724) 1992–94	Toys (894) 1986–87	Toys (894) 1992–94	Footwear (851) 1986–87	Footwear (851) 1992–94	Travel goods (831) 1986–87	Travel goods (831) 1992–94
China	100	100	100	100	100	100	100	100	100	100
India	79	94	..	..	..	..	..	..	..	..
Indonesia	42	47	..	..	..	..	..	..	..	..
Malaysia	31	27	279	226	..	11	..	..	..	..
Philippines	76	84	..	90	22	19	85	20	14	19
Thailand	58	40	..	74	18	27	128	36	32	27

Note: Numbers in parentheses are Standard International Trade Classification codes.
Source: UN COMTRADE database.

TABLE 4.4

World export market shares for Hong Kong and mainland China, 1992 and 2020
(percent)

Industry	1992 Hong Kong	1992 China	Change to 2020 Hong Kong	Change to 2020 China
Primary agriculture	0.0	4.7	0.1	−2.5
Processed food	0.4	3.4	0.7	−2.8
Oil and gas	0.0	1.4	0.0	−1.4
Other energy	0.0	2.6	0.0	−2.0
Other natural resources	0.1	2.5	1.9	−2.0
Textiles	2.4	8.1	−1.5	3.1
Wearing apparel	9.8	16.9	−8.9	25.8
Light manufactures	1.4	6.5	2.2	10.4
Heavy manufactures	0.4	1.8	0.4	2.7
Transport, machinery and equipment	0.9	1.5	1.3	13.1
Utility, housing, and services	21.6	0.0	−6.9	0.0
Other services	2.0	2.1	−1.2	1.8
Total	1.3	3.0	0.2	6.8

Source: GTAP database, version 3; World Bank staff projections.

manufacturing (chemicals, rubber, plastics, paper, iron and steel, and nonferrous metal) China gains some market share (4 percent), but industrial countries maintain their dominance. More intraindustry trade patterns can also be expected in China's overall trade, as can greater intensity of trade with its neighbors (box 4.2)

Terms of trade effects. A rapid rise in exports of labor-intensive manufactures by China will depress their prices, while China's increased imports of primary products and machinery and transport equipment will raise their prices. Thus the resulting terms of trade changes for any country will depend on its relative spe-

BOX 4.2

China's rising trade with East Asia

China's trade links with its neighboring lower-middle-income economies have remained weak despite the rapid development of the Chinese economy in the past twenty years. Trade links have strengthened only with the richer economies in the region (the Republic of Korea and Taiwan, China) and with North America outside the region. But baseline projections to 2020 suggest that the weak export orientation of China toward developing countries in East Asia is set to change dramatically. The proportion of Chinese exports going to four East Asian countries (Indonesia, Malaysia, Philippines, and Thailand) will quadruple. This is mainly because of the increasing specialization of Chinese exports in the light manufactures and capital goods imported by its neighbors.

China's export orientation to East Asia
(percentage of total merchandise exports)

	1978	1989	1992	1995	2020
Hong Kong (China)	26.0	41.4	43.9	24.2	14.8
Other newly industrialized economies	2.5	3.2	6.0	8.9	10.9
Korea, Rep. of	0.0	0.0	2.9	4.5	5.0
Singapore	2.5	3.2	2.4	2.4	2.4
Taiwan (China)	0.0	0.0	0.8	2.1	3.6
Others	3.3	2.5	2.6	3.7	9.8
Thailand	0.7	0.9	1.0	1.2	4.2
Malaysia	1.7	0.7	0.8	0.9	2.7
Indonesia	0.0	0.4	0.6	1.0	2.1
Philippines	0.9	0.5	0.2	0.7	0.8

Source: IMF, *Directions in Trade Statistics,* various years; World Bank staff projections.

cialization among these products (table 4.5). Expected losses are significant for East Asian and South Asian developing countries. Downward pressure on relative prices for products in which developing countries compete intensely are confirmed by the historical record—

TABLE 4.5

Changes in world market shares and terms of trade for developing countries, 1992–2020

Country	Changes in world market share for all sectors (percentage points)	Cumulative change in terms of trade (percent)
Republic of Korea, Taiwan (China), Malaysia	1 to 3	–3 to –9
Asian developing countries (excluding newly industrialized economies)	0.3 to 2.8	–8 to –13
Sub-Saharan Africa, Latin America, rest of world	–0.2 to 0.1	–3 to 4

Source: World Bank staff projections.

2.5 percent a year in clothing and 1.5 percent a year in footwear in the United States since the mid-1980s. However, the volume gains from growing world trade and rising market shares more than offset such losses.

Effects of a slowly integrating China

What are the implications for growth, trade, and welfare in the rest of the world if China's integration proceeds more slowly than expected? Slower integration could occur because of higher trade barriers put up by China and by the rest of the world. Such a reversal from baseline trade policies is assumed to have its main impact by slowing productivity growth in China. This is assumed to slow Chinese GDP growth from 7 percent (in the baseline) to about 4 percent a year. The adverse effects of such a slow down could be large.

The welfare cost to China of a reversal from baseline trade policies and slower integration dwarfs the adverse impact on the rest of the world (although that impact is underestimated in the model). China forgoes welfare gains equivalent to $1.2 trillion (1992 U.S. dollars); the cost to the rest of the world is $107 billion. Regions not directly competing with China (high- and low-income countries) lose most when China's growth slows because their exports are lower, their terms of trade suffer, and their domestic resource allocation worsens. China's closest competitors (countries in Asia and Brazil) collectively gain about $5 billion a year, as terms of trade improve and exports are higher for labor-intensive goods.

The volume of world exports is 7.5 percent lower than the baseline in 2020, however. Losses to the rest of the world from slower growth in China would be over $100 billion a year in perpetuity (equal to 0.2 percent of their GDP in 2020). But this is a gross underestimate because the comparative static model does not account for a number of dynamic effects: a slower rate of capital accumulation (and lower returns to capital) because of reduced world trade and slower growth, lower efficiency in the rest of the world as a result of lower competition from China, and reduced investment opportunities in China.

Notes

1. The Global Trade Analysis Project provides data and a standard model for multicountry general equilibrium analysis. The project is maintained at Purdue University and is guided and supported by an international consortium that includes the World Bank, OECD, World Trade Organization, and others. The description and applications of the model are provided in Hertel (1997). The standard model is a comparative static model and does not include dynamic effects, such as those from endogenous capital accumulation. It also does not include the effects of scale economies or imperfect competition. As result the standard model's results are only a fraction of those that can be expected in reality. This chapter's results rely on a baseline simulation provided by Professor Thomas Hertel and associates at Purdue.

2. See Muscatelli (1996) on the impact of newly industrialized economies on industrial economies.

3. A few economists such as Wood (1994, 1995) consider trade with developing countries to be an important factor explaining the deteriorating situation of unskilled workers in jobs and relative wages. Wood's argument is that new labor-saving technologies are themselves encouraged by low-wage competition, and that the existence of large low-wage competitors depresses wages even though trade is small.

4. The revealed comparative advantage index is the ratio of the share of a product in a country's total exports to the share of that product in world exports. An index greater than one for a product indicates a comparative advantage in exporting that product.

References

Ahuja, V., and D. Filmer. 1995. "Educational Attainment in Developing Countries." Policy Research Working Paper 1489. World Bank, Washington D.C.

Boltho, A., U. Dadush, D. He, and S. Otsubo. 1994. "China's Emergence: Prospects, Opportunities, and Challenges." Policy Research Working Paper 1339. World Bank, Washington, D.C.

Chow P., M. Kellman, and Y. Shachmurove. 1994. "East Asian NIC Manufactured Intra-Industry Trade, 1965-1990." *Journal of Asian Economies* 5(3): 335–348.

Hertel, Thomas, ed. 1997. *Global Trade Analysis: Modelling and Applications.* Cambridge: Cambridge University Press.

Krugman, P. 1994. "Does Third World Growth Hurt First World Prosperity?" *Harvard Business Review* (July–August): 113–21.

Martin, Will. 1993. "The Fallacy of Composition and Developing Country Exports of Manufactures." *The World Economy* (March): 159–72.

Muscatelli, A. 1996. "Flexibility, Structural Change, and the Global Economy." University of Glasgow Discussion Paper in Economics 9601. University of Glasgow.

OECD (Organisation for Economic Co-operation and Development). 1995. *The OECD Jobs Study: Investment, Productivity and Employment*. Paris.

———.1996a. *Trade and Labor Standards*. Paris.

———.1996b. "Vertical Intra-Industry Trade Between China and OECD Countries." OECD Development Center Technical Paper 114. Paris.

Rodrigo, G. Chris, and Will Martin. 1996. "Can the World Trading System Accommodate More East Asian–Style Exporters?" World Bank, Washington, D.C.

Wood, A. 1994. *North-South Trade, Employment, and Inequality: Changing Fortunes in a Skill-Driven World*. Oxford: Clarendon Press.

———. 1995. "How Trade Hurt Unskilled Workers." *Journal of Economic Perspectives* 9 (3): 57–80.

World Bank. 1997. *Global Economic Prospects and the Developing Countries 1997*. Washington, D.C.

Yeats, Alexander. 1991. *China's Foreign Trade and Comparative Advantage*. World Bank Discussion Paper 141. Washington, D.C.

Changes in China's Tariffs, 1992–2005

ANNEX TABLE A.1

Tariffs and offers for the GTAP sectors

Sector	Applied 1992	Applied 1995	Applied 1996	WTO 2005
Paddy rice	−35.2	0.0	0.0	0.0
Wheat	−12.9	0.0	0.0	0.0
Coal	15.0	12.0	6.0	6.0
Oil	2.0	1.5	1.5	1.5
Gas	33.6	29.7	12.5	12.5
Beverages and tobacco	124.1	74.0	66.6	53.7
Textiles	62.4	55.0	30.6	26.9
Wearing apparel	79.6	75.0	39.7	28.7
Pulp, paper	23.1	22.3	16.7	11.5
Petroleum and coal products	12.3	10.6	9.0	9.0
Chemicals, rubber, and plastics	22.2	20.4	14.4	12.5
Nonmetallic mineral products	35.7	28.8	21.7	17.7
Primary ferrous metals	12.8	12.6	9.5	8.3
Nonferrous metals	13.6	12.7	10.3	6.5
Fabricated metal products	40.2	38.3	22.6	17.6
Transport industries	57.9	43.9	38.3	31.7
Machinery and equipment	26.1	22.7	16.2	13.5
Other manufacturing	66.3	58.2	42.2	29.0
Weighted average	31.2	28.1	19.8	16.2
Unweighted average	39.1	35.6	23.4	17.5

Note: The GTAP sectors are defined in terms of the International Standard Industrial Classification System.
Source: World Bank staff estimates.

ANNEX TABLE A.2
China's average tariffs by country/region
(percent)

Country or region	Applied 1992	Applied 1995	Applied 1996	WTO 2005
Australia	11.6	18.7	16.8	13.0
Japan	38.1	32.5	22.8	19.3
Korea, Rep. of	44.1	39.6	24.1	20.6
ASEAN	22.0	23.9	17.9	13.9
Hong Kong, China	44.8	39.7	25.5	20.3
Taiwan, China	41.4	37.5	23.4	19.2
South Asia	19.9	19.3	11.8	9.4
Canada	5.9	54.0	51.4	31.1
United States	22.1	23.4	19.3	15.0
Latin America and the Caribbean	12.6	17.8	15.1	10.5
European Union	33.3	29.3	22.4	18.5
Austria, Finland, Sweden	20.5	21.4	16.5	11.6
Iceland, Norway, Switzerland	21.4	18.8	13.2	10.8
Central European Associate	20.3	18.7	15.4	13.2
Countries of the former Soviet Union	13.4	17.7	16.2	9.5
Middle East and North Africa	15.6	12.4	9.8	9.2
Sub-Saharan Africa	14.8	19.5	16.0	9.3
Rest of world	20.6	21.3	16.0	12.2
Weighted average	31.2	28.1	19.8	16.2
Unweighted average	39.1	35.6	23.4	17.5

Source: World Bank staff estimates.

ANNEX TABLE A.3
The declining standard deviation of tariffs
(percent)

Sector	Applied 1992	Applied 1995	Applied 1996	WTO 2005
Paddy rice	0.00	0.00	0.00	0.00
Wheat	0.00	0.00	0.00	0.00
Coal	3.67	2.95	1.55	1.55
Gas	10.08	9.43	3.42	3.30
Beverages and tobacco	50.18	30.79	19.38	21.24
Textiles	23.17	20.66	10.46	8.82
Wearing apparel	11.45	9.83	4.26	2.57
Pulp, paper	19.35	17.60	15.31	8.69
Petroleum and coal products	8.74	8.07	3.71	3.54
Chemicals, rubber, and plastics	19.32	14.40	10.61	7.37
Nonmetallic mineral products	20.01	18.45	14.38	8.67
Primary ferrous metals	7.44	7.37	4.81	4.36
Nonferrous metals	16.60	15.56	10.02	5.82
Fabricated metal products	16.45	15.54	7.78	7.69
Transport industries	50.18	37.05	27.96	24.38
Machinery and equipment	21.94	17.68	11.85	9.50
Other manufacturing	22.70	20.29	14.12	8.19
Total	27.35	23.79	16.68	11.49

Note: The GTAP sectors are defined in terms of the International Standard Industrial Classification System.
Source: World Bank staff estimates.

ANNEX TABLE A.4
Overview of simulations of China's tariff liberalization

Variable	
Baseline	Projections to 2005. Shocks to factor endowments, population, and total factor productivity in agriculture. Implicit total factor productivity in other sectors derived from GDP targets. No trade reforms, but 1992 import subsidies to agricultural products removed, so these sectors are exposed to world market prices.
UR	Incorporates the effects of the Uruguay Round agreement (excluding China and Taiwan, China).
C96	Shows the impact of trade liberalization in China as set out in the 1996 APEC tariff announcement.
WTO	Further liberalization in China as set out in the second Chinese offer to the WTO for 2005.
CTW	Same as for WTO but with abolition of Multi-fibre Arrangement quotas on textiles and wearing apparel for China, Hong Kong (China), and Taiwan (China).

Source: World Bank staff estimates.